PROCEEDINGS OF THE

FIRST LUNAR INTERNATIONAL LABORATORY (LIL) SYMPOSIUM

RESEARCH IN GEOSCIENCES AND ASTRONOMY

ORGANIZED BY THE
INTERNATIONAL ACADEMY OF ASTRONAUTICS
AT THE XVITH INTERNATIONAL ASTRONAUTICAL CONGRESS
ATHENS, 16 SEPTEMBER 1965
AND DEDICATED TO THE TWENTIETH ANNIVERSARY
OF UNESCO

EDITED BY

F. J. MALINA

INTERNATIONAL ACADEMY OF ASTRONAUTICS
PARIS

WITH 43 FIGURES

Springer-Verlag Wien GmbH

1966

ISBN 978-3-7091-8155-3   ISBN 978-3-7091-8153-9 (eBook)
DOI 10.1007/978-3-7091-8153-9

©1966 BY SPRINGER-VERLAG WIEN
Originally published by Springer Vienna in 1966.
Softcover reprint of the hardcover 1st edition 1966

LIBRARY OF CONGRESS CATALOG CARD NUMBER 66-24187

TITLE No. 9173

Preface

The Lunar International Laboratory (LIL) project of the International Academy of Astronautics was begun upon the proposal of the editor at the First Special Meeting of the Academy at Stockholm on 16 August 1960. The late THEODORE VON KÁRMÁN, first President of the Academy, appointed the following members of the LIL Committee: Prof. N. BONEFF (Bulgaria), Prof. M. FLORKIN (Belgium), Mr. A. G. HALEY (U. S. A.), Prof. Sir BERNARD LOVELL (U. K.) (Vice-Chairman), Prof. L. MALAVARD (France), Dr. F. J. MALINA (U. S. A.) (Chairman), Prof. H. OBERTH (German Federal Republic), Dr. W. H. PICKERING (U. S. A.), Prof. E. SÄNGER (German Federal Republic), Prof. L. I. SEDOV (U. S. S. R.), Prof. L. SPITZER, JR. (U. S. A.), Dr. H. STRUGHOLD (U. S. A.), Prof. H. C. UREY (U. S. A.) and himself. Since 1960 the following additional members were appointed to the Committee: Mr. A. C. CLARKE (U. K.), Prof. A. DOLLFUS (France), Prof. Z. KOPAL (U. K.), Dr. S. F. SINGER (U. S. A.), Prof. N. M. SISSAKIAN (U. S. S. R.) and Prof. F. ZWICKY (Switzerland).

The Academy authorized the Committee to study the technical problems related to the construction of a manned research laboratory on the Moon and the feasibility of carrying out its construction, and to consider the fields of research which would initially be undertaken.

In 1964, at Warsaw, the Committee organized a Discussion Panel during the 15th I. A. F. International Astronautical Congress to review, publicly, in a general way, the scientific and technical aspects of such a laboratory. The Panel felt that work in a lunar manned research center might begin in the decade 1975–85. The Panel also outlined broadly research that might be carried out uniquely on the Moon.

On the basis of the Panel discussions, the LIL Committee made plans for a series of annual symposia, each symposium to be devoted to one or more scientific fields.

The present volume contains the papers in the fields of the geosciences, astronomy and astrophysics presented at the First Lunar International Laboratory Symposium. The Symposium was held during the 16th I. A. F. Congress at Athens on 16 September 1965, and was dedicated to the 20th anniversary of Unesco.

The session on geosciences was held under the chairmanship of Prof. K. Y. KONDRATIEV, Rector of the University of Leningrad. Prof. F. ZWICKY of the California Institute of Technology chaired the session on astronomy and astrophysics. The sessions were well attended, and the organizers of the symposium were gratified by the interest and discussion the papers provoked. In order to reduce the delay in bringing out this volume, it was decided not to include a summary of the discussions.

The Second LIL Symposium, devoted to life sciences research and lunar medicine, will be held during the 17th I. A. F. Congress at Madrid in October 1966.

On behalf of the LIL Committee and the contributors to the symposium I wish to express great appreciation to Dr. C. S. DRAPER, President of the Academy, and to the Academy Secretariat for support and aid in organizing the first symposium; to the International Astronautical Federation and to the Hellenic Astronautical Society for making it possible for the symposium to be held in Athens; and to Springer-Verlag for their cooperation in publishing the proceedings.

Boulogne sur Seine, France, 26 January 1966

Frank J. Malina

Contents

Contents

Research and System Requirements for a Lunar Scientific Laboratory

By

C. William Henderson[1] and Grady L. Mitcham[2]

(With 7 Figures)

Abstract — Résumé — Резюме

Research and System Requirements for a Lunar Scientific Laboratory. In order to realistically determine the desirability of a Lunar International Laboratory and to establish the time frame of its implementation three basic criteria must be established: the types of research to be conducted from a LIL, the personnel and equipment support required of and for a LIL, and the most promising concept of hardware for the laboratory.

Though many scientific disciplines could utilize a lunar station for basic and applied research, the following list represents the most attractive options:

Geosciences: Geology; Geophysics; Geochemistry.

Astronomy: Optical; Radio; Astrophysics.

Biosciences: Exobiology — search and analysis; Biological effects of lunar environment on earth specimens; Biomedicine (applied research); Ecological Systems (applied research).

Physics: Lunar atmosphere and radiation; Geomagnetic field; Lunar materials properties; Effects of lunar environment on earth materials; Beam investigations.

Earth Oriented: Synoptic meteorology; Synoptic oceanography.

Basic shelter, life support and power are, of course, prime requisites for any lunar station. In addition, a laboratory should provide equipment and space for working. It should be emphasized, however, that because of the extreme weight limitations, space and equipment will be minimal, necessitating unique and perhaps yet unknown research techniques especially adapted to a lunar station.

We must also assume that long stay time capacity should be an inherent quality of a LIL. In addition there must be a reasonable degree of comfort and recreation for the crew. It might be pointed out that general housekeeping will occupy from 30 to 40% of the working day, thereby limiting the available time for research. As an adjunct to base operations as well as research, some degree of lunar surface mobility will be required in addition to any vehicles required for geological traverses.

Support required for the station would necessitate sufficient and reasonably scheduled transportation systems to and from earth. Perhaps the most important of all support will be adequate earth based data analysis personnel and facilities (special centers and universities) to handle the flood of information from the moon. Unless this capability exists to process the lunar data, an extensive scientific operation on the moon could not be justified.

How a lunar laboratory will be constructed must be determined in the light of operational problems on the moon. Extensive physical labor for construction in the

---

[1] Manned Lunar Missions Directorate, Advanced Manned Missions Program, NASA Office of Manned Space Flight, Washington, D. C., U.S.A.

[2] Lunar Program Manager, Advanced Space Systems Development, Aeroscape Division, The Boeing Company, Seattle, Washington, U.S.A.

lunar environment will be precluded by difficult motions in a space suit, the unreliability of workmanship and the extreme cost of a lunar man hour. This would imply that construction, or even assembly is not desirable. Therefore, the LIL should be modular, composed of different elements assembled and completely equipped on earth. In this concept only minimal labor would be required to connect the elements together, and the system would have great flexibility for future additions or modifications.

By the mid 1970's, man could be technically capable of landing payloads of 25,000 pounds or more with a single launch vehicle. This size module is quite adequate to provide any reasonable element of a LIL, such as large shelters, roving vehicles, 200 kw nuclear power plants, telescopes and vast amounts of supplies.

Matériel nécessaire au fonctionnement et aux travaux de recherche d'un Laboratoire Scientifique Lunaire. Pour pouvoir déterminer avec réalisme l'opportunité d'un Laboratoire Lunaire International et établir le calendrier de son installation, il importe de préciser: le type de recherches effectuées à partir du LIL (Lunar International Laboratory); le personnel et l'infrastructure nécessaires dans le laboratoire et sur la terre; la conception la plus favorable pour la construction du laboratoire.

Quoique de nombreuses disciplines scientifiques puissent utiliser une station lunaire pour la recherche fondamentale ou appliquée, la liste suivante représente le choix le plus séduisant:

Sciences du globe lunaire: géologie lunaire; sélénophysique; sélénochimie.

Astronomie: astronomie optique; radio-astronomie; astrophysique.

Sciences de la vie: exobiologie (la vie hors de la terre) recherche et analyse; effets biologiques de l'environnement lunaire sur les espèces terrestres; biomédecine (recherche appliquée); systèmes écologiques (recherche appliquée).

Physique: atmosphère et rayonnement lunaires; champ magnétique naturel; propriétés des matériaux lunaires; effets de l'environnement lunaire sur les matériaux terrestres; recherches à l'aide de faisceaux.

Observation de la terre: météorologie synoptique; océanographie synoptique.

L'abri servant de base, les moyens de vie et les sources d'énergie sont évidemment les exigences fondamentales de toute station lunaire. Un laboratoire doit, de plus, fournir l'équipement et l'espace nécessaires pour travailler. Il faut cependant insister sur le fait que, par suite de l'extrême limitation de poids, l'espace et l'équipment seront réduits au minimum, nécessitant des techniques de recherches particulières et peut-être encore inconnues, spécialement adaptées à une station lunaire.

Nous devons aussi admettre qu'une des qualités inhérentes au LIL doit être la capacité de durer longtemps. On doit y trouver, de plus, confort et distraction pour l'équipage à un degré raisonnable. Il faut signaler que les besognes d'entretien et d'intendance occuperont de 30 à 40% de la journée de travail, limitant ainsi le temps disponible pour la recherche. Pour les opérations de la base aussi bien que pour la recherche, une certaine mobilité est nécessaire sur la surface de la lune en plus des véhicules destinés aux expéditions géologiques.

La station nécessitera pour fonctionnement un système de transport venant de la terre et y retournant, suffisant et raisonnablement ordonné dans le temps. La partie la plus importante de toute l'infrastructure sera peut-être, un personnel et des moyens d'analyse de données, situés sur terre, et adéquats (centres spéciaux, universités) pour traiter le flot d'informations venant de la lune. Si cette possibilité de traiter les donnés lunaires n'existe pas, une opération scientifique d'envergure sur la lune ne se justifie pas.

La façon dont un laboratoire lunaire sera construit doit être déterminée en fonction des problèmes opérationnels sur la lune. Un travail physique de construction considérable dans l'environnement lunaire est exclu par la difficulté de mouvement dans le costume spatial, la qualité incertaine du travail exécuté dans ces conditions et le coût extrêmement élevé d'une heure de travail sur la lune. C'est pourquoi le LIL doit être fait éléments, les divers éléments étant assemblés et complètement équipés sur terre. Suivant cette conception, le travail nécessaire pour réunir les éléments entre eux sera minime et le système présentera une grande souplesse permettant des additions et des modifications ultérieures.

Vers 1975 on serait techniquement capable d'envoyer des charge utiles de 12 tonnes et plus à l'aide d'un simple véhicule de lancement. Cette dimension de charge set tout à fait suffisante pour fournir au LIL tout élément utile tel que: grands abris, véhicules autonomes, sources d'énergie nucléaire de 200 kW et des fournitures en grande quantité.

Научно-исследовательская работа и требования, предъявляемые к научной лаборатории на Луне. Для определения с реальных позиций желательности создания Международной Лунной Лаборатории (МЛЛ) и выработки графика осуществления этого мероприятия следует исходить из трех основных критериев: виды исследований, которые будут проводиться МЛЛ, персонал и оборудование как для самой лаборатории, так и для обеспечения ее функционирования, наиболее многообещающие конструкции тяжелого оборудования и помещений для лаборатории.

Хотя лунная станция будет использоваться для исследований теоретического и прикладного характера самыми различными научными дисциплинами, наибольший интерес она, вероятно, будет представлять для исследований в следующих областях:

Науки о Земле: геология; геофизика; геохимия.

Астрономия: оптическая; радиоастрономия; астрофизика.

Биологические науки: экзобиология — поиски и анализ; биологический эффект лунной среды на земные виды; биомедицина (прикладные исследования); экологические системы (прикладные исследования).

Физика: лунная атмосфера и радиация; геомагнитное поле; свойства лунных материалов; эффект лунной среды на земные материалы; исследования излучений.

Исследования, касающиеся условий на Земле: синоптическая метеорология; синоптическая океанография.

Первоочередными потребностями лунной станции являются потребность в помещении для жилья, средствах жизнеобеспечения и энергии. Кроме того, лаборатория должна располагать оборудованием и местом для работы. Однако, следует подчеркнуть, что из-за строгих весовых ограничений, размеры помещения и количество оборудования будут минимальными, потребуются уникальные, возможно еще неизвестные, методы исследований, специально приспособленные для условий лунной станции.

Следует также предположить, что одним из важнейших свойств МЛЛ должна быть способность работать в течение длительного времени. Кроме того, она должна обеспечивать достаточную степень комфорта и возможностей для отдыха экипажу. Можно указать, что обычная домашняя хозяйственная работа будет занимать примерно 30—40 проц. рабочего времени, что будет ограничивать время для исследовательской работы. В дополнение к работам общего характера и исследованиям потребуются средства для перемещения в каких-то пределах по поверхности Луны, сверх любых средств транспорта, необходимых для выполнения геологических разрезов.

Для обеспечения работы станции потребуется организовать достаточно эффективные и работающие по разумному графику системы транспорта между Землей и Луной. Вероятно, одним из важнейших средств обеспечения работы станции должны быть люди и средства, осуществляющие обработку данных на Земле (специальные центры и университеты), которые могли бы обеспечить обработку потока данных, поступающих с Луны. При отсутствии таких возможностей для обработки данных, поступающих с Луны вряд ли будет оправдана крупная научная экспедиция на Луну.

Вопрос о конструкции лунной лаборатории должен решаться с учетом проблем выполнения работы на Луне. Следует исключить конструкции, требующие больших затрат физического труда при монтаже, из-за трудности передвижения в скафандре, отсутствия необходимой рабочей силы и чрезвычайно высокой стоимости каждого часа работы экипажа на Луне. Из этого следует, что строительство, или даже монтаж конструкций, нежелательны. Поэтому МЛЛ должна быть модульной конструкцией, состоящей из отдельных элементов, смонтированных и оборудованных на Земле. При такой конструкции потребуется лишь минимум труда для соединения вместе отдельных элементов, кроме того, конструкция обеспечит необходимую гибкость на случай дальнейшего расширения или модификации станции.

К середине семидесятых годов будет технически возможно доставлять на Луну с помощью одного космического корабля полезный груз порядка 25.000 фунтов и более.

Модуль такого размера достаточен в качестве элемента МЛЛ, такого, как жилое помещение, вездеход, атомная электростанция мощностью в 200 кв, телескоп или большое количество различных припасов.

A program of scientific and exploratory work on the moon is the logical sequel to the first manned lunar landings. However, the exact course of lunar surface exploration and exploitation cannot yet be predicted since uncertainties exist in many areas, including detailed data on the lunar surface characteristics and environment, the level of funding of lunar operations, and the relative importance of possible lunar missions. It is probable that many potential lunar operations will undoubtedly require support by large facilities on the lunar surface. Although the size, lifetime, and purpose of these facilities may vary with the specific missions, all must perform the fundamental functions of providing shelter, life support, power, communications, and surface mobility.

Before we can realistically define the requirements for a lunar scientific laboratory, we must first establish two preliminary sets of criteria. The first set of criteria will define the basic objectives and the extent of science that might be conducted in a lunar laboratory; and hopefully it will indicate the potential value to mankind of the scientific information to be gained. The second set of criteria will establish promising concepts for lunar habitability, and should define the cost of laboratories in terms of manpower and material resources as well as a realistic scheduling for laboratory buildup. From these two sets of criteria we can then establish the specific experiments and the specific lunar exploration and mission equipments and personnel to conduct the missions.

At the present time, these preliminary criteria are in their early stages of development. However, we can already envision many exciting areas of science which would, indeed, be desirable to carry out at a lunar laboratory. There have also been conceived several systems which could support man for his lunar scientific efforts.

It should be pointed out that most of mankind's ventures on Earth into the unknown are gradual in nature, evolving from simple cautious investigations into bolder and more extensive operations. Likewise, it is expected that lunar science will evolve from simple investigations during the early manned landings into complex experiments requiring large permanent stations. Therefore, a lunar laboratory will most probably grow from a small temporary station into a large permanent base, building up gradually as scientific requirements, engineering technology, monetary resources and international cooperative efforts permit.

Of the potential objectives for lunar science that are envisioned, those oriented toward the direct study of the Moon and embracing the many fields of the geosciences may be obvious. However, the scientific research potential of a lunar laboratory is certainly more extensive than investigations concerned with the Moon itself. Consequently, the major objectives of lunar science may be categorized as follows:

1. Investigations concerned with the nature, environment, history, and origin of the Moon;

2. Investigations for which a lunar surface site is used as a platform for studying space itself, the Sun, planets, stars and galaxies;

3. Investigation utilizing a lunar platform for studies of the Earth, the Earth-Moon system, and cislunar space; and

4. Investigations which make use of the lunar site and the lunar environment to carry out fundamental experimentation not possible on Earth or from satellites.

The operational and general logistic requirements which these investigations would impose on an advanced lunar system may be quite divergent. Early geoscience investigations will be principally of a geological and geophysical nature and, for the most part, will involve reconnaissance and mapping with limited fixed-laboratory activities. Such systems will probably consist of manned roving vehicles operating either alone or from a single small shelter. Later geoscientific activities will be more heavily geophysically oriented, and will need greater fixed-laboratory capacity, greater capability for deep drilling, and long time periods for deep seismic surveys.

To accomplish early geosciences investigations, several (perhaps three) small laboratories 400 to 700 miles apart on the lunar surface would be desirable. One of these sites may ideally provide the beginnings of a more permanent station with greater capability than the others. In this manner, the need for a more sophisticated laboratory at a particular site could be established before committing hardware for a large facility.

It is difficult to detail the geoscientific requirements for a large permanent lunar laboratory at this time, since geoscience activities, by comparison with optical astronomy and radioastronomy, are rather mobile in nature. A two-to-six-man laboratory, capable of crew-occupation periods of three to six months, that can be reoccupied for short periods appears to satisfy the requirements for early geoscientific support. As mentioned previously, these would be required at a number of different lunar-surface locations. Two traverse vehicles would be extremely desirable, as would be a fixed-base capability of drilling to depths of approximately 1,000 feet.

Lunar resource exploitation may be one important exception to the apparent lack of requirements for a large permanent lunar base in support of geoscientific activities. While this may not be strictly a basic scientific operation, it is closely related to the geosciences area. The existence and utilization of lunar resources could have major effects on base system concepts for support of other lunar scientific operations.

Optical astronomy investigations will have a major influence on any system visualized for support of scientific investigations during the later phases of a long-term lunar laboratory. A probable minimum size optical telescope with which astronomers could perform investigations of major scientific significance on the lunar surface is approximately a 40-inch telescope.

If optical astronomy experiments are to be performed on the Moon of a substantial nature and of such a character that the scientific return will be commensurate with the effort invested, then a large reflecting telescope may be desirable, such as a 100-inch diffraction limited instrument. Such a telescope, in the postulated seeing environment of the lunar surface, would represent a significant advance in capability.

Optical astronomy investigations can be introduced during lunar scientific missions at three general levels of activity or equipment; (1) small wide-field or high resolution telescopes, possibly of 12-inch diameter, (2) intermediate size (40-inch) telescopic system, and (3) large size (100-inch or more) telescopic system.

The first small telescope should be included in an early lunar station because the experience gained from the use of this small instrument is considered necessary for the development of lunar astronomy techniques.

The use of a 40-inch telescope from the lunar surface in multi-hour exposures during later scientific missions will permit the study of faint objects at least as well as the 200-inch telescope on earth. The principal stellar applications would be (1) in long-exposure work in those parts of the spectrum where earth-based telescopes are not capable of probing, and (2) in following up, in detail, the myriad problems which initial orbital system telescopic surveys most assuredly will uncover. The 40-inch telescope will also provide experience for eventual use of a much larger instrument.

The 100-inch telescope which would appear desirable for use during the later phases of lunar scientific missions would have, in theory, eighty times the capability of the 40-inch telescope to detect and resolve signals. No large telescope should be without smaller supporting instruments. The most effective approach is to utilize the smaller instrument until sufficient data is in hand to justify the use of the larger telescope. The 100-inch instrument, in addition to performing investigations unique to its capabilities, could then be used to extend the investigations initiated by smaller lunar surface instruments, telescopes in orbit, and terrestrial telescopes.

Radioastronomy investigations will be of major importance during all phases of extended lunar exploration. No other scientific area is expected to yield as much broad scientific information for the effort expended. In general, the early radioastronomy will have two purposes: to produce scientific information of value, as well as identify the requirements for more sophisticated experiments, such as those that may be associated with placing a radioastronomy observatory on the farside of the Moon. Such an installation may provide a truly unique location for sensitive, long-term, detailed radioastronomy studies of the universe. Because of the very great potential of a listening post on the farside of the Moon, shielded from terrestrial interference and enjoying a wide window in the radio spectrum, preparatory work toward this goal is of primary significance in lunar scientific missions.

Early radioastronomy investigations might involve the emplacement of simple wire antennas, beacons, etc. These activities initially would require a base crew for placement or deployment, calibration, and initial operation. After the initial activities, the operation could be automatic or be remotely controlled except for periodic inspection or maintenance.

The steerable dish antennas, required for the later investigations, would demand construction capabilities. However, these requirements are not unique, and it is expected that the construction equipment required for placement of any base system will be adequate for the radioastronomy equipment.

The most desirable base system arrangement would appear to be a manned permanent base near the lunar limb, where it is always in sight of the Earth. The actual radioastronomy sensors or antennas could then be located up to 200 miles beyond the limb or at the maximum range of a ground reconnaissance lunar roving vehicle to account for lunar librations. This would insure shielding from electromagnetic interference from the Earth.

In addition to radioastronomy, some potential high frequency and particle radiation experiments might be of major importance for all phases of lunar exploration. A better understanding of space radiation, particularly solar radiation, is of critical importance for crew safety during lunar surface operations and other space missions. The radiation investigations that could be performed during early periods are independent of base size, crew activities, or base deployment location, and the equipment, power, volume, and weight requirements are modest. The equipment could be mostly automatic, requiring little or no attention after initial set-up and operation, except for periodic checks and readjustment. Some radiation-astronomy investigations may be performed during the later phases of lunar exploration that would involve somewhat unique operations; however, these investigations have system support requirements related to optical astronomy, and in themselves do not affect base-system concepts, provided optical astronomy observatory operations are present.

Three radiation experiments are considered to be extremely attractive investigations to be conducted from the lunar surface. They are: (1) high-energy galactic particle scattering studies; (2) solar-wind flow past the Moon including plasma-sheath observations; and (3) measurement of the ultra-violet and X-ray spectrums of the Sun, stars, and galaxies. It may be especially desirable to conduct the solar-wind and magnetic-field experiments simultaneously at widely separated lunar-surface locations, 500 to 700 miles apart, with the solar-plasma probes and the magnetometer in close proximity to each other.

Physics experiments are of major importance during all phases of lunar exploration for reasons of crew safety (environmental considerations) and future base operations. The more important investigations are related to basic scientific research with potential application to engineering problems, and would encompass lunar surface chemistry, exploitation of lunar resources, and material behavior. The possible physics investigations can be divided into three categories: (1) basic physics experiments, (2) studies of the lunar atmosphere, and (3) lunar and Earth materials research in the lunar environment.

At this time, it does not appear that the requirements for system support of the physics scientific investigations will have a significant effect on lunar laboratory concepts or configurations. An exception may be the exploitation of lunar materials. However, in the event it is feasible to mine and refine lunar materials, the possible required activities are so varied and diverse that any attempt to define lunar-base support requirements should await more information regarding the materials and characteristics of the lunar surface.

Biology investigations concerning aspects of exobiology are expected to be of primary importance in the early lunar exploration phase. These investigations will include the search for fossils, chemicals of biological significance, and living forms. The requirements for support of the exobiology investigations are small, and quite similar to the geoscience investigations, sharing much of the equipment, with the possible exception of extremely exacting sterilization and decontamination equipment. Nevertheless, any discovery of life or fossils on the Moon would be a major scientific breakthrough.

Later biological investigations to evaluate ecological system components will be of major importance for the future growth of the laboratory. These investigations will require special support requirements of the base system.

Biological laboratory volume requirements progress from a few cubic feet to a small room-size laboratory, but a firm size requirement is not possible at this time.

The scientific crew in the biology laboratory might include two men; one in addition to a medical doctor. The second man should be a biologist, and the part-time assistance of an electrical engineer would also be desirable. A large laboratory may require a biochemist and a physiologist in addition to part-time aid from other disciplines.

Biomedical investigations will be of major importance for all phases of the scientific exploration of the Moon. Short-term crew safety considerations will be of immediate importance followed by later long-term crew performance investigations that will have significant bearing on the long-term lunar operations. The investigations foreseen to be of interest (mainly detailed crew activity observations), tend to be independent of base size or phase. Requirements for base support are negligible except for the recommendation that one of the crew should be a medical doctor and that he have adequate instrumentation. With the exception of providing an M.D., there are no special crew requirements. In addition to his medical duties, the M.D. could be expected to assist in laboratory analysis activities for geoscience and bioscience.

The investigations of interest will be studies of lunar gravity effects, general health and safety, mechanical efficiency of man in reduced gravity, crew work capability, and psychological effects.

Specific experiments are difficult to formulate. Presently there are no well defined biomedical problems that are known to be best for study in the lunar environment. The later experiments that may be recommended will depend upon the results of present investigations for manned Earth orbital space stations and the initial manned lunar landings.

Meteorology observations of the Earth from the lunar surface may provide significant scientific data, but are, for the most part, of secondary importance. Exceptions are two investigations which may be of considerable interest:

(1) atmospheric heat balance, and (2) reflectivity and albedo. Meteorological satellites provide a closer platform for collecting synoptic meteorological data than the Moon; therefore, the proposed lunar investigations are research oriented.

The Moon would be an excellent platform for planetary meteorology investigations including those of general atmospheric circulation, temperature, and composition studies and the measurement of gross atmospheric properties.

Meteorology investigations will not affect the lunar-base systems concept appreciably or its location on the earthward side of the Moon, since no unique requirements for base crew personnel are foreseen and the equipment required for the meteorology investigations are of a general purpose nature, including spectrographs and a small telescope.

Oceanography investigations from the lunar surface, which may have considerable scientific merit, are: (1) heat flux measurements of the ocean surfaces and cloud cover involving the measurement of sea surface temperatures over large areas for determination of thermal content; (2) sequential photography to obtain information and measurements on sea surface glitter, contrasts, colors, and ice pack formation; and (3) sea surface height measurement for measurement of slope and ellipticity of the geoid.

The oceanography investigations are closely related to meteorology, and do not have unique requirements that will affect the base system concept or any laboratory location on the earthward side of the Moon. With the exception of a ranging device (perhaps laser) to provide an accuracy of 10 meters or better for tidal and geoid measurements, no large development activity should be required to accommodare the instruments. Special crew requirements are not apparent except that the oceanographic observations should be made by a crewman oriented in geoscience or bioscience with appropriate additional training in oceanography.

As can be seen there are, indeed, many worthy and important scientific objectives that could be accomplished on the surface of the Moon. However, we must consider how these objectives could be supported in terms of mission equipment to house the personnel and their instruments. It is reasonable to assume that the type of system most suited to support this spectrum of scientific objectives should be easily adapted to both limited and extensive experiments, small and large personnel compliments and to varying degrees of stay time, with permanency being a desired goal.

It is difficult to imagine lunar laboratories of the 1970 to 1980 decade as being similar to the large complexes we have on Earth. Since any component of a lunar laboratory must be limited in size and weight to the payload capacity of its launch vehicle, a lunar laboratory should be composed of an assemblage of payload packages. However, extensive requirements for human assembly on the lunar surface should be minimized, since physical labor for construction in the lunar environment will be precluded by difficult motions in a space suit. Also fabrication and test could be accomplished far more reliably and more economically in our factories on Earth, and the manpower on the Moon should be used to the maximum for investigation. This would imply that construction and even complex assembly on the Moon is not desirable. Therefore, the laboratory might be modular in form, composed of different large elements, each assembled and completely equipped on Earth. In this concept only minimal labor would be required to connect the elements together, and the system would be adaptable to future additions or modifications.

This type of system would consist of a family of prefabricated modules that could be arranged on or below the lunar surface in a variety of arrays to support a range of missions. The example laboratory to be discussed would facilitate expansion of installations both to accommodate increased lunar scientific activity, and to decrease dependence on resupply from Earth.

It would cover a range of base sizes varying from 3 to 18 men, and having different occupancy durations of from 3 to 24 months. The basic shelter modules are designed for deployment on the lunar surface without the necessity of subsurface excavation, and are sized to be landed by potential lunar landing vehicles of the 1975 time period. The landed payload capacity of the lunar landing vehicle has been assumed to be in excess of 25,000 pounds.

Fig. 1. Base Model 1 for housing three men for a period of up to three months (NASA MT 65-9965, 9-1-65)

Fig. 1 shows a design that could house 3 men for a period of up to 3 months. The basic module has an interior volume of 2,800 cubic feet (80 cubic meters) which, in actuality, could accommodate a crew of up to 6 men. About half of this volume is occupied by stores and equipment. This shelter is designed as a rigid structure consisting of two concentric pressure vessels. The inner vessel is a cylinder with ellipsoidal ends, and the outer vessel is a toroid. Part of the toroid consists of an airlock between the lunar environment and the inner vessel. Inside the shelter would be packaged food, water, atmosphere, and all equipment required to operate the shelter, and perform experiments. A lunar surface vehicle could be delivered with the shelter as a part of the same payload, thereby providing surface mobility to the scientists and astronauts. The shelter and roving vehicle combination could be delivered unmanned to the Moon, and should be capable of remaining unattended for several months before occupancy. This concept might be referred to as Base Number 1.

Fig. 2 is a top view of the mockup of this shelter showing an entire 6 man crew. This figure also shows the location of sleeping quarters, living area, emergency

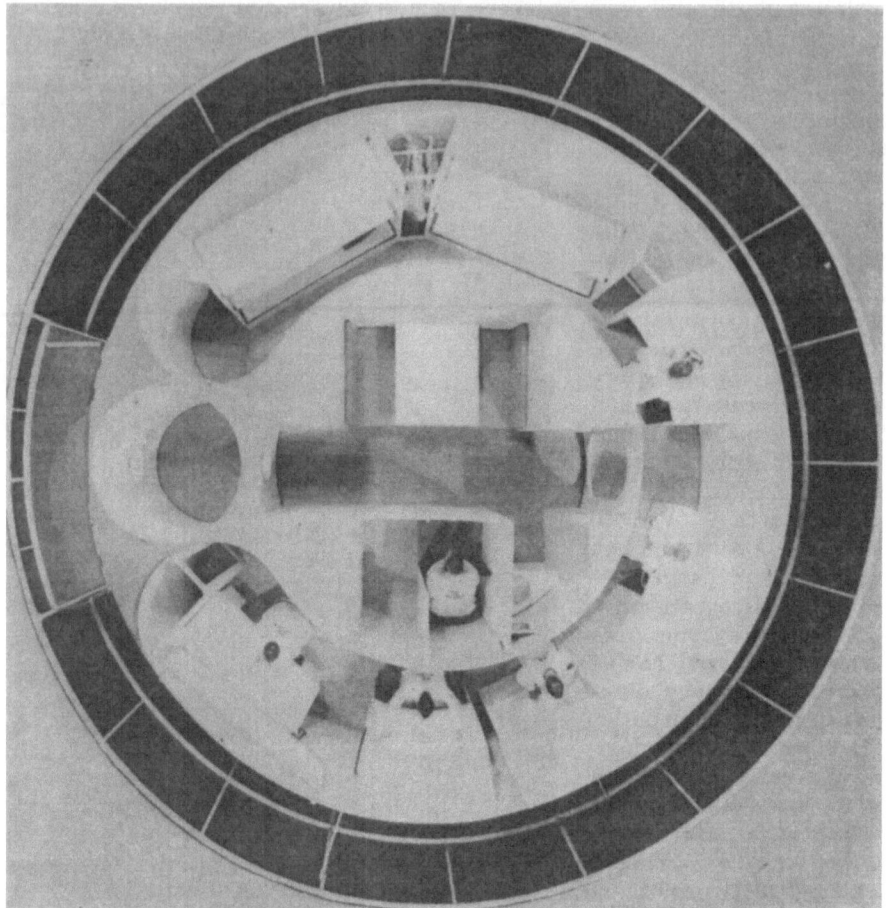

Fig. 2. Top view of mockup of Base No. 1 (NASA MT 65-9963, 9-1-65)

medical area, pressure suit storage, maintenance bench, scientific work area, commander's quarters, and duty stations. Electric power would be provided by liquid hydrogen — liquid oxygen fuel cells.

A typical lunar surface vehicle is shown in Fig. 3. This vehicle would be capable of operating over a wide range of possible lunar terrain, conducting missions up to a total of 250 miles (400 km), with a stay time capability of 14 days with two astronaut-scientists. The overall weight would be in the order of 7,000 pounds (3200 Kg). The cabin is pressurized to approximately one-half atmosphere with 50% oxygen, which would allow a shirt sleeve operating environment. An airlock is provided for ingress and egress. Liquid hydrogen and liquid oxygen fuel cells would provide the basic power to the electrical motors which drive each individual wheel. In the example design shown, the aft unit is attached by an elastic frame unit to the forward wheel unit to provide greater obstacle climbing capabilities. The range of this vehicle could be extended greatly by the addition of selfpowered fuel modules, or by the substitution of radioisotopes as the fuel source.

Fig. 4 shows the same basic shelter module as in Base 1 with an increased crew occupancy of 6 men and a possible duration of 6 months. This Base Model 2

Fig. 3. Lunar surface vehicle for missions of up to 400 km

would provide the scientists with more stay time and scientific instrument capability. In Base Model 2 the electrical power requirements of the increased operation might demand a small nuclear reactor. Such a power plant would be a separate module and would require remote placement as well as integral radiation shielding (or utilize lunar materials for shielding). The shelter of Base Model 2 must be protected with soil filling a caisson around its perimeter. The use of lunar fill could result in substantial savings in weight and costs if the fill is found to be readily available but, of course, would require human effort as well as soil moving equipment.

Fig. 5 depicts a base buildup for 12 men for 12 months. Two basic shelter modules are utilized here and could be connected if desired by an airlock system. This base would definitely require nuclear power.

Fig. 6 shows the base buildup to 18 men for 24 months. Again, the same basic shelter modules are utilized, and in the background are shown cargo versions of the lunar landing vehicle. Larger laboratories could be built upon this basic design, perhaps having special modules for specific types of investigations. An example might be housing for a large telescope. Also, extensive electrical power could be made available by the addition of nuclear reactor modules.

The various base models are built up with primary payloads, each containing a crew shelter, roving vehicle, and stores of expendables sufficient for Base Model 1 operation; and by supplementary payloads providing additional shelters, base equipment and stores of expendables as the bases become larger. The expendables contained in the primary payloads are directly integrated with the various applicable subsystems, hence their tankage constitutes the prime base storage facilities. All other expendables, either those delivered as a part of base establishment or those for base duration extension (resupply) must depend on some auxiliary means to permit their delivery to the using system tanks.

Fig. 4. Base Model 2 for housing six men for a period of up to six months (NASA MT 65-9964, 9-1-65)

Though the operational requirements for a lunar laboratory will not be discussed extensively in this paper, it should be pointed out that not all the time spent by scientists on the Moon can be devoted to science. Fig. 7 shows an estimate of the average time spent by the crew in various activities of a lunar mission. These activity allocations would naturally differ from day to day and from person to person.

As would be expected, approximately 50% of the time (12 hours per day) would be spent for personal activities. The remaining 50% would be devoted to productive work. Of the average work day, about 60% can be used for science. This percentage may be optimistic, since less than 40% of the working time is available for science in research activities now being conducted in the Antarctic.

The figure shown is for a 6 man — 6 months lunar mission. Missions of lesser magnitude require that a greater percentage of personnel time be devoted to base support activities, whereas larger missions require somewhat less time. This is readily understood when it is recognized, as an example, that a small life support system generally has a similar number of components as a large system, and therefore, requires approximately the same amount of maintenance despite the size of the crew. Also, large missions can afford specialization of crew members, thereby permitting increased efficiency.

As we all know, the build-up and maintenance of a permanent lunar laboratory in this century will draw very heavily upon the resources of Earth; in terms of materials, technology and labor. We, the scientists and engineers advocating lunar exploration are not the only ones who must pay for this expensive venture;

Fig. 5. Base Model 3 for housing twelve men for a period of up to twelve months
(NASA MT 65-9960, 9-1-65)

Fig. 6. Base Model 4 for housing eighteen men for a period of up to twenty four months
(NASA MT 65-9961, 9-1-65)

rather, it is the people and governments of the world. Unless stimulated, their interests in the Moon may wane rapidly after the first few lunar landings, especially if there is strong competition for funds from other space projects or from other requirements on Earth.

It is our obligation to define a meaningful lunar laboratory program which will extract the maximum scientific return for the investment of resources. Our scientific methods may have to change radically in order to achieve this goal of efficiency; perhaps by using the mantime on the Moon only for the extraction of data, leaving the analysis of this information to the far less costly manpower on Earth. Most probably, leading scientists will not be, themselves, on the lunar surface, but rather will be at Earthbased television consoles directing and guiding

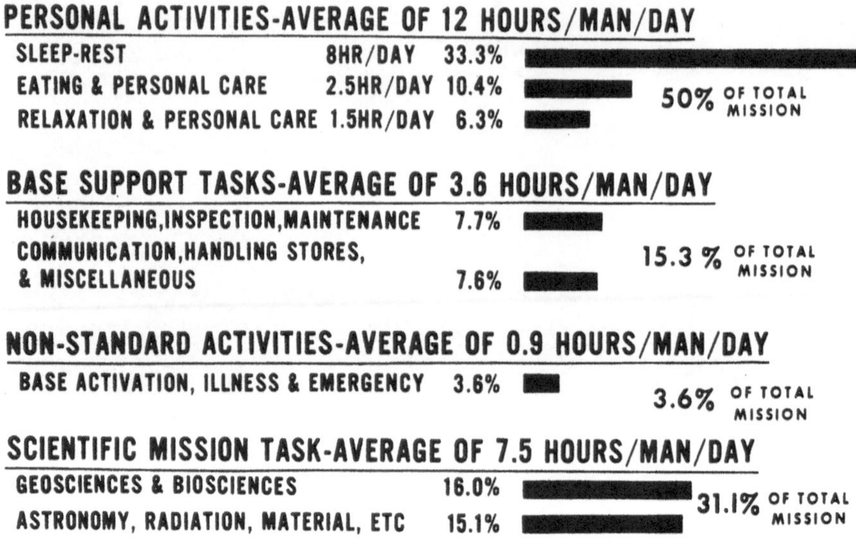

PERSONAL ACTIVITIES-AVERAGE OF 12 HOURS/MAN/DAY

SLEEP-REST	8HR/DAY	33.3%	
EATING & PERSONAL CARE	2.5HR/DAY	10.4%	50% OF TOTAL MISSION
RELAXATION & PERSONAL CARE	1.5HR/DAY	6.3%	

BASE SUPPORT TASKS-AVERAGE OF 3.6 HOURS/MAN/DAY

HOUSEKEEPING,INSPECTION,MAINTENANCE	7.7%	
COMMUNICATION,HANDLING STORES, & MISCELLANEOUS	7.6%	15.3 % OF TOTAL MISSION

NON-STANDARD ACTIVITIES-AVERAGE OF 0.9 HOURS/MAN/DAY

BASE ACTIVATION, ILLNESS & EMERGENCY	3.6%	3.6% OF TOTAL MISSION

SCIENTIFIC MISSION TASK-AVERAGE OF 7.5 HOURS/MAN/DAY

GEOSCIENCES & BIOSCIENCES	16.0%	31.1% OF TOTAL MISSION
ASTRONOMY, RADIATION, MATERIAL, ETC	15.1%	

Fig. 7. Estimated crew time allocations for a six man—six months mission (NASA MT 65-5993)

their counterparts on the Moon. Concurrently, a vast staff of technicians and scientists may be working in Earth laboratories on previously received data. The methods of science on the Moon will undoubtedly change from our present concept of individual investigation to that of group participation in which large numbers of scientists each has his specific role in the overall analysis of new and significant discoveries. Just as engineering has evolved from the individual inventors at the turn of the century into vast teams of men and women who now build our spacecraft today, so too must scientific analysis evolve for maximum return.

In light of the fact that many scientific manhours will be required on Earth, and that each expensive launch operation must yield us the greatest amount of scientific data for that particular point in time, we must address a seldom considered problem: How will this data be transformed into knowledge?

If we return to Earth by telemetry or by actual physical retrieval, vast amounts of data, we must be able to assimilate this information in a timely and efficient manner. If there are many missions to the Moon, each returning information requiring many years to reduce to usuable form, lunar science would probably lose public support. In essence we must not allow lunar scientific data to accumulate in storage vaults for lack of personnel and facilities to examine and interpret it. A major task of this group, here, will be to define, in terms of scientists, technicians, equipment and facilities, the requirements on Earth to analyze lunar scientific data in a timely fashion.

Perhaps the most pressing task facing those who advocate lunar science in depth, is the requirement to firmly establish a program that will be compatible with future space capabilities at the time those capabilities exist. Technology is moving rapidly, so it is not unreasonable to predict that a small 3 man lunar base could be in existence in 10 years. It would be extremely desirable that this base and future base designs reflect the needs of science.

To incorporate scientific requirements into future space plans it is necessary to establish the scientific goals, especially those which are peculiar to the Moon or which could be accomplished most advantageously on the Moon. From these goals a program can then be established.

It is reasonable to expect that lunar science will be evolutionary, forming from the results of early experiments and growing with the advance of space technology capabilities. This evolution must be anticipated for the orderly planning of a step program. If the specific experiments of the future cannot be defined early, at least the gross demands upon a laboratory system must be predicted in terms of performance, time and costs. In this manner the long lead time required to produce equipment and spacecraft for scientific needs will not emerge as a surprising and perhaps hindering factor in the program.

In brief, planning of a lunar science program must start now in order that science can and will be able to participate fully in this, one of the most exciting adventures of mankind.

References

1. G. L. Mitcham, Earth Moon Transportation Systems. Society of Automotive Engineers, New York, 1963.
2. T. C. Evans, Lunar Flight Programs, Vol. 18, p. 480. American Astronautical Society, Western Periodicals Co., North Hollywood, Calif., 1964.
3. P. D. Lowman and D. A. Beattie, Lunar Flight Programs, Vol. 18, p. 453. American Astronautical Society, Western Periodicals Co., North Hollywood, Calif., 1964.
4. J. R. Lopik, R. A. Geyer and C. Crowe, Early Mission Experiments and Lunar Resources Exploration. 3rd Annual Meeting, Working Group on Extraterrestrial Resources, Air Force Missile Development Center, Alamogordo, N.M., 1964.
5. C. W. Henderson, Extended Lunar Exploration. 11th Annual Meeting American Astronautical Society, Chicago 1965.
6. F. J. Malina, Lunar International Laboratory. Space Flight 7, 155 (1965).
7. F. J. Malina, Report of the Lunar International Laboratory Discussion Panel, Warsaw. Astronaut. Acta 11, 123 (1965).

Commentary on the Paper of C. W. Henderson and G. L. Mitcham

By

Edward Z. Gray[1]

The authors, in their paper "Research and System Requirements for a Lunar Scientific Laboratory", have devoted their attention to a long duration laboratory on the Moon which would be of major international interest. This certainly appears to be within our grasp technologically, should such an objective be established. However, since there are some precursory activities required prior to the establishment of such a station, I would like to discuss with you the lunar exploration planning that is aimed at the time between the period of the initial manned lunar landing and that of a long duration lunar scientific station.

There is no question in the minds of most people that a long duration laboratory will be deployed on the Moon. However, there are many questions still to be resolved regarding its timing, its size, its purpose, its location or locations, the number required, and the system on Earth needed to support such an endeavor. The expense attendant to operate such a laboratory sytem will be great by comparison to our present concept of Earth-based laboratories. A most important consideration, therefore, bearing on the decision for establishment of a lunar laboratory, will be the results from the lunar exploration program which precedes it.

Recently, specific exploration goals were established by leading scientists in the fields of geology, geophysics, biosciences, particles and fields, petrology, selenography, and astronomy. These goals ranged from the initial measurements to be made by the astronaut with the 250 lbs. of scientific equipment provided in the early Apollo flights to those requireng expanded Earth-Moon transportation systems.

I can report to you some of these general objectives which will have a major bearing on shaping our lunar exploration plans and the systems required to support them. The general objectives for lunar exploration include the following:

1. A comprehensive geological and geophysical investigation of the entire Moon utilizing multiple sensors from orbit combined with surface traverses by trained men over selected areas of greatest interest.

2. A detailed search for selected samples of the lunar surface which are to be returned to Earth for complete analysis. These specimens will also include those having the greatest interest to the biologists in their search for life or indications of previous life.

3. The establishment of emplaced instruments which will gather data on the changing phenomena that exist on the Moon such as temperature variations, micrometeoroid flux, seismic activity, etc.

---

[1] Director, Advanced Manned Missions Program, National Aeronautics and Space Administration, Washington, D. C., U.S.A.

4. Measurement of the physical characteristics of the Moon: i.e., its gravity and magnetic characteristics, its atmosphere, radio-activity, etc.

5. The observation of the universe from the Moon to determine the value of the Moon as a base for extension of our astronomical activities here on Earth and in Earth orbit.

6. Investigation of the behavior of various physical phenomena such as friction of materials in the environment of low gravity and high vacuum, the action of fluids in a 1/6 g gravity field, the formation of surface features resulting from the deposition of various size particles on the lunar surface, etc.

7. Development of the capability to operate in the environment of the Moon with the combination of low gravity, high temperature differentials, foreign materials, low atmospheric pressure, and unknown surface characteristics.

8. Investigate the resources available on the Moon and the means to use them for supporting lunar exploration and/or planetary missions.

The NASA for these recommended scientific objectives is determining the instruments required, identifying the supporting equipment such as environment, astronaut time, etc. and evaluating the various systems to support these operations.

Feasibility and conceptual design studies are being conducted in each of the following topics which in many cases will result in more than one experiment:

1. Geological Investigations:
   (a) Field geology
   (b) Field resources

2. Geophysical Investigations:
   (a) Drills, shallow and deep
   (b) Emplaced scientific instruments
   (c) Subsurface probes
   (d) Active seismology
   (e) Gravity measurements
   (f) Magnetic field measurements
   (g) Heat flow measurements
   (h) General physical properties of representative samples
   (i) Photometric properties
   (j) Magnetic properties
   (k) Velocity and propagation studies of sound, elastic and shock waves in the lunar environment
   (l) Thermoluminescence
   (m) Passive seismic investigation
   (n) Effects of solar proton bombardment in ultra-high vacuum

3. Geochemical Investigations:
   (a) Support of laboratory investigations on returned samples
   (b) Mass spectrometry
   (c) Other absorption spectroscopy
   (d) Gamma-ray spectroscopy
   (e) X-ray diffractometry
   (f) X-ray fluorescence analysis
   (g) Gamma-ray spectrometry
   (h) Microscope instrumentation
   (i) Chemical reactions in the absence of gases
   (j) Neutron analysis

(k) Lunar gas and other emanation detection
(l)   Vapor pressure determination of compounds in lunar environment
(m) Validity of gas laws of chemistry in a lunar environment

4. Engineering Investigations:

(a) Surface behavior of particles in ultra-high vacuum

5. Astronomy and Astrophysics:

(a) Optical astronomy experiment package
(b) Photography of nearby interstellar media
(c) Radio astronomy investigations
(d) Solar wind/X-ray astronomy

6. Remote sensing of Lunar Surface:

(a) Gamma radiation
(b) X-ray fluorescence
(c) UV fluorescence
(d) Visual photography
(e) Infra-red mapping
(f) Passive microwave investigations
(g) Radar mapping
(h) UHF reflectivity
(i) Absorption spectroscopy
(j) Gravity gradient measurements
(k) Micrometeorite measurements

From the above listing, which is yet incomplete, one can see that a broad scope of activity is being considered during the early phases of lunar exploration.

The systems available for the conduct of these experiments include the Surveyor, the Lunar Orbiter, and the Extended Apollo Systems and their possible follow-ons. Each of the experiment objectives, individually or as a collected group of experiments, is being matched against the capabilities of these systems to determine the most economic manner of conducting a total exploration program.

The initial manned lunar landings of the Apollo program will concentrate on determining the feasibility of operations on the lunar surface. The astronauts will examine the physical characteristics of the surface materials, the homogeneity of these materials, their bearing capacity, the problems of movement over the surface, the selection of samples for return to Earth, and other characteristics of the lunar environment. They will also emplace a 150 lb. module containing instruments which will continue to measure a few selected environment conditions after the astronauts leave.

It is expected that the initial landings would be followed by one or more manned lunar orbital surveys containing multiple sensors as described above to select the most promising locations for follow-on landings. By means of high resolution photography, correlated with radar, UV, infrared, and other sensors, it is expected the areas of greatest interest will be revealed even if there is a surface layer of dust covering the moon. The collelation of Earth based thermal gradient measurements with the visual photography of crater locations indicates the feasibility of such orbital flights for the preliminary selection of surface exploration sites.

Studies have indicated that soft landing probes can be carried on these manned lunar orbital survey flights. It appears attractive to consider the use of a modified Surveyor spacecraft for such use. When carried to lunar orbit by the

Apollo system, its payload capability is greatly increased over what it can carry when Earth launched with an Atlas-Centaur. Such probes will permit seismological measurements at several locations of the Moon's surface to greater depths than possible with early manned landings.

Soon after the initial lunar landings, the Apollo capability will be increased both in payload capacity, stay time duration and surface mobility. This is one of the objectives of the Extended Apollo System.

By landing an unmanned LEM prior to and in conjunction with manned Apollo landings, up to one and a half tons of experimental instrumentation and supporting equipment can be placed on the surface of the Moon for manned exploration. Studies have shown that this will permit the use of a wheeled surface vehicle to extend man's radius of geologic and geophysic operation from a few hundreds of meters to at least 10 kilometers. It will also permit deployment of seismological instrumentation, transportation of surface samples, and extension of the duration of the astronauts extravehicular operation. Preliminary analyses have shown that all the scientific instrumentation for lunar exploration fall within the capability of the Extended Apollo Exploration system individually. However, when the supporting equipment and the astronaut time available are also included in planning a typical lunar mission, then selectivity of each mission objective must be practiced. There is a trade-off possible between stay time on the Moon and the equipment to be used.

For longer duration surface missions, a direct landing logistic system will be required. This can be developed using either the Saturn IB/Centaur or the Saturn V launch vehicle. In either case an added landing stage must be developed. If the Saturn V logistic system is developed, then a long term lunar surface labotarory will become feasible, since more than a 15-ton payload can be placed on the Moon.

There is an alternative approach to a lunar laboratory, however, depending upon its purpose. It is possible to deploy a laboratory in orbit around the Moon from which continuous close observations can be made of any surface changes. It would provide far greater coverage than a stationary laboratory on the surface. From an orbital laboratory, short duration excursions could be made to the lunar surface for close-up investigations. Such a laboratory could be a logical extension of an Earth orbital laboratory program should such a system be developed.

Our studies have shown that the long term manned missions having greatest realism of scientific return at this time are optical or radio astronomy stations. The Moon's lack of background interference and its stability are promising characteristics for such observations. Almost all the other investigations discussed appear to involve dispersed equipment, short duration measurements, emplaced unmanned instruments, or surface mobility. It is, therefore, believed that an astronomical observatory will form the basic element of any long term lunar laboratory. We have, therefore, included in our missions planning early investigation of the desirability of optical, radio, and X-ray observations from the Moon. Only after some actual experience there will we able to judge the value of a lunar astronomy program.

In conclusion, I would like to summarize by stating that in the time period immediately following the initial Apollo lunar landing, the United States will embark on a broad program of scientific investigation of the Moon. From the results of these investigations will come the requirements for a scientific laboratory on the Moon. We believe the program will contribute much toward man's understanding of the universe in which we live.

Recherche géologique lunaire

Par

J. H. Focas[1] et A. Dollfus[2]

Résumé — Abstract — Резюме

Recherche géologique lunaire. On résume les données disponibles sur les propriétés thermiques et optiques du sol lunaire établies à l'aide d'observations radar et radio-astronomiques, de mesures de polarisation, photométriques, colorimétriques, et de luminescence et de recherches comparatives en laboratoire sur des susbtances terrestres, perrmettant certaines conclusions sur la nature du sol lunaire.

On essaie de prévoir quelles seront nos connaissances sur la morphologie de la surface lunaire et l'origine des formations géologiques au début du programme LIL en tenant compte des techniques d'observations téléscopiques ou à l'aide d'engins spatiaux du type Orbiter ou Surveyor ou autres, mis en œuvre par différents pays. On rapproche les caractères morphologiques de certaines formations terrestres avec ceux des formations lunaires photographiées par les engins spatiaux permettant: la discrimination entre les éléments volcaniques et météoriques; l'étude de l'association des cônes centraux, des plissements, des failles, etc. avec des formations supposées analogues aux calderas; l'étude de la quantité et de la production de la poussière recouvrant le sol lunaire, de sa distribution sélénographique et de la consistance du sol lunaire, l'association d'alignements de cratères avec de grands cratères parents, etc.; ces données permettent une classification géologique des formations.

En utilisant les résultats précédents, il est probable que les géologues concernés par le LIL auront à leur disposition: a) les cartes complètes de la topographie lunaire; b) des cartes géologiques classant par régions, les épanchements et les recouvrements par matériaux éjectés; c) des cartes des anomalies thermiques, de l'importance des échos radar, des propriétés électro-magnétiques du sol, etc., donnant les premières interprétations pour la préparation du travail.

La recherche géologique directe pourrait alors porter sur les objets suivants: au cours des déplacements sur le sol lunaire, les documents photographiques et les mesures telles que les déterminations de la consistance et des températures du sol, des radio-activités, et l'analyse ultérieure d'échantillons prélevés, permettront un examen détaillé des différences dans les propriétés du sol. La prise et l'analyse d'échantillons obtenus en profondeur au moyen de forages, perforés en différentes profondeurs permettront d'étudier les propriétés électriques et thermiques du sol, la radio-activité, la vitesse de propagation du son, le champ magnétique, la conductivité du sol, le flux de chaleur non solaire, etc. Une attention particulière sera portée aux formations superficielles telles que les cratères récents, leurs intérieurs, leurs rebords, les matériaux éjectés lors de leur formation, etc., les cratères secondaires, les cratères anciens fortement erodés, les pitons centraux, les champs de lave, les failles, les dépressions du sol, etc.

La composition du sol et l'étude des émissions gazeuses peuvent avoir des conséquences pratiques.

---

[1] Observatoire de Meudon, France, Ancien Astronome de l'Observatoire National d'Athènes, Grèce.

[2] Observatoire de Meudon, France.

Lunar Geological Research. We have summed up the data available on the thermic and optical properties of the lunar soil, obtained by means of radar and radio-astronomic observations, by measurements of polarization, photometric and colorimetric measurements, measurements of luminescence and comparative laboratory research on terrestrial substances, leading to certain conclusions about the nature of the lunar soil.

We endeavoured to foresee how much we shall know about the morphology of the lunar surface and the origin of its geological formations at the beginning of a LIL programme, with the use of telescopic observation techniques and the aid of spacecraft of the Orbiter, Surveyor and other types operated by different countries. We have compared the morphological characteristics of certain terrestrial formations with those of the lunar formations photographed by spacecraft. This has enabled us to: discriminate between volcanic and meteoric formations; study the association between the central cones, folds, faults, etc... and the formations supposedly resembling calderas; examine the quantity and production process of the dust covering the Moon's surface, its selenographic distribution and the consistence of the lunar soil; study the association between alignments of craters and the major parent crater; and so on. All these data furnish a basis for geological classification of the lunar formations.

Using the above-listed data, it is probable that LIL geologists will have at their disposal: a) a complete map of the lunar topography; b) geological maps classifying by regions the flows and successive layers of substances ejected; c) maps showing thermic anomalies, scale or radar echoes, electro-magnetic properties of the soil, etc., with preliminary interpretations for the preparation of work.

Direct geological research could then concentrate on the following: travelling about the Moon collecting photographic documentation and measurements to be used later, in conjunction with data on soil consistence and soil temperatures, and radioactivity, and with the results of the analysis of samples collected, for purposes of making a detailed examination of the properties of the soil. The analysis of depth samples obtained by borings and holes dug at various depths will furnish information for study of the electrical and thermic properties of the soil, the radioactivity, the speed of propagation of sound, the magnetic field, the conductivity of the soil, the flux of non-solar heat, etc. Special attention will be paid to surface formations such as recent craters—their cavities, their rims, the substances ejected at the time of their formation, etc.; secondary craters; ancient, strongly eroded craters; central peaks; lava fields; depressions in the soil, etc.

Investigation of the soil composition and of the gaseous emanations may have practical consequences.

Геологические исследования на Луне. Собрать и обработать имеющиеся данные о термических и оптических свойствах лунного грунта, полученные путем радиолокационных и радиоастрономических наблюдений, поляризационных, фотометрических, калориметрических и люминесцентных измерений, сравнительных лабораторных исследований земных веществ, дающих возможность прийти к определенным заключениям относительно характера грунтов на Луне.

Нужно попытаться предвидеть, какими знаниями о морфологии поверхности Луны и о происхождении геологических формаций мы будем располагать к началу программы МЛЛ, с учетом методов оптических наблюдений или наблюдений с помощью космических исследовательских станций, таких как Обритер, Сарвейор и другие, создаваемых различными странами. Сопоставив морфологический характер некоторых земных формаций с характером лунных формацией, сфотографированных космическими станциями, можно провести различие между вулканическими и метеоритными элементами. Изучение ассоциаций центральных конусов, складок, трещин и т.д. одновременно с изучением предположительно аналогичных им кальдеров; изучение количества и продуцирования пыли, покрывающей лунный грунт, связей рядов кратеров с большими „родительскими" кратерами и т.д. — все это позволит провести геологическую классификацию формаций.

Вполне вероятно, что геологи, заинтересованные в МЛЛ, использовав результаты всех вышеперечисленных работ, будут располагать: а) полными картами лунной топо-

графии; б) геологическими картами с показом районов, залитых или покрытых выбро-
шенными породами; в) картами термических аномалий, силы радиолокационного эхо,
электромагнитных свойств грунтов и т.д., дающими первые интерпретации для подготовки
к работе.

После этого непосредственные геологические исследования могут проводиться в
следующем порядке: при передвижении по поверхности Луны — фотографирование и
выполнение таких измерений, как определение консистенции и температуры грунта,
радиоактивности, взятие проб для последующего анализа; все это позволит детально
изучить различия в свойствах грунта. Взятие, путем бурения, проб из глубины грунта, их
анализ, бурение в различных местах и на разную глубину скважин позволит изучить
электрические и термические свойства грунта, радиоактивность, скорость распространения
звука, магнитное поле, проводимость грунта, тепловой поток не-солнечного происхож-
дения и т.д. Особое внимание нужно будет уделить поверхностным образованиям, таким
как свежие кратеры, их внутренние и наружные поверхности, материалы, выброшенные
при их формировании, и т.д., вторичные кратеры, старые кратеры, сильно подвергшиеся
эрозии, центральные пики, поля лавы, трещины, опустившиеся участки грунта и т.д.

Изучение состава грунта и выделения газов может дать результаты практического
характера.

Des travaux classiques ont permis de tenter certaines interprétations sur la
nature du sol lunaire et sur l'évolution probable des formations qui le recouvrent.

Le vol des sondes spatiales Lunik III et Zond ont fourni des informations sur
les lignes générales du relief et la distribution des grandes formations de l'hémisphère invisible de la Lune.

Les photographies prises par Rangers VII, VIII et IX ont montré la structure
fine de régions choises, représentatives de l'origine des formations y contenues.

On prévoit la continuation des programmes pour l'obtention de photographies
à très haute résolution de vastes régions de la surface lunaire et l'utilisation
d'instruments permettant l'étude des propriétés physiques du sol.

Selon les programmes Surveyor et Luna, des instruments seront posés sur
des régions choises de la Lune photographier la surface et pour l'étude de la
pesanteur, des radiations, du magnétisme, de la séismicité et de la minéralogie de
notre satellite par le recueil d'échantillons du sol.

Selon le programme Orbiter, de vastes régions de la lune seront photographiées
depuis un satellite artificiel, avec une résolution très élevée.

Selon le programme Prospector, un engin sera posé sur la Lune, ayant la
capacité de se mouvoir à la surface et prendre des mesures de toutes sortes à
l'aide d'instruments transportés.

Le programme Apollo prévoit l'atterrissage sur la Lune d'équipes humaines.

Ces programmes et d'autres non encore annoncés par d'autres pays, devraient
permettre l'installation d'un laboratoire lunaire International.

a) Etat actuel de nos connaissances en géologie lunaire

Nos connaissances sur les propriétés physiques du sol lunaire, acquises par
les techniques modernes, peuvent se résumer comme suit:

1. Des agglomérations de particules de dimensions inférieures à 1 micron,
déposées sur le sol selon un mécanisme insuffisamment connu, créent la diffusion
et la polarisation de la lumière observée depuis la Terre.

Ses propriétés thermiques étudiées au cours des lunaisons, confirment l'existence d'une couche de telle poudre recouvrant le sol en toutes ses parties, d'une
épaisseur d'au moins 1 millimètre et d'une densité réduite.

Au-dessous de cette couche superficielle, la densité augmente avec la profondeur, la cohésion des matériaux étant de beaucoup inférieure à celle des rochers. Celle-ci est atteinte à une profondeur de quelques mètres au moins.

La réflexion des échos radar est beaucoup plus forte sur les formations récentes que sur des formations anciennes. Ceci indique une rugosité et une cohésion supérieures sur les formations plus récentes. Les formations anciennes sont aplanies par le transport de matière poudreuse résultant des impacts, selon des procédés complexes dans lesquels pourraient intervenir des phénomènes électrostatiques.

La radiation thermique de la Lune a été étudiée sur des régions non illuminées par le Soleil. Elle montre que des matériaux récents ont une température supérieure à celle des matériaux anciens. On pourrait supposer l'existence de sources d'énergie indépendantes, comme les volcans, des coulées de lave fraîche, un flux de chaleur interne facilitée par la conductivité locale des matériaux de la surface, déclenchée par l'action d'impacts. L'émission thermique des mers, qui présente de grandes variétés, pourrait être justifiée en partie par la fuite de gaz chauds ou froids sous l'action de séismes. L'émission rougeâtre observée dans la région d'Aristarque et le cas d'Alphonsus pourraient également être justifiée par un procédé de dégazement.

Les études polarimétriques entre 0,5 et 1,05 microns de longueur d'onde sur des régions lunaires choisies, et des mesures comparatives effectuées sur des substances terrestres en laboratoire, montrent que le sol lunaire pourrait être recouvert d'une poudre constituée de grains sombres et très absorbants. La proportion de lumière polarisée varie en sens inverse de la longueur d'onde utilisée et de la brillance de la région. Ces propriétés polarisantes caractérisent des laves volcaniques en état poudreux, ou certaines poudres claires assombries sous l'action d'un bombardement de protons. Un tel bombardement est produit sur la Lune par le vent solaire.

Des techniques spectrométriques très sensibles montrent l'existence d'un excès de lumière diffusée de 2 à 10% sur certaines régions de la surface lunaire, attribuables à un effet de luminiscence produit par irradiation de ces région lors des éruptions chromosphériques solaires. Des expériences effectuées en laboratoire à l'aide d'un accélérateur à protons, sur des échantillons de météorites pulvérisés, montrent une forte luminescence de ces derniers dans le domaine spectral de 6.900 Å et une plus faible vers 4.000 Å.

Des mesures d'émission infra-rouge, donnent pour le point sub-solaire une température de 125° C et pour le centre de l'hémisphère non éclairé une température d'environ −150° C. Des mesures effectuées pendant les éclipses lunaires montrent une chute brusque de la température due à la conductivité thermique très faible du sol lunaire. Cette propriété caractérise les matériaux poreux. Pourtant, certaines régions montrent température plus élevée et par conséquent une conductivité supérieure.

2. L'origine des formations lunaires, en général, fait encore l'objet d'interprétations contradictoires ou de spéculations.

Les documents classiques montrant la morphologie du sol lunaire avec une résolution d'environ 500 mètres; ils ont déjà permis d'essayer de remonter dans l'histoire de la Lune et de se former une idée sur l'évolution de sa surface, en adoptant cinq systèmes de stratification successifs: le Pré-Imbrien, le Imbrien, le Procellarien, l'Eratosthénien et le Copernicain, chacun représentant une période géologique. D'après une autre théorie, on pourrait distinguer quatre périodes géologiques: L'archéosélène, la Protosélène, la Mésosélène et la Téléosélène. La première et la deuxième périodes pourraient s'étendre de l'état protoplanétaire, jusqu'au début de la contraction de la croûte lunaire. La Mésosélène est caractérisée par la contraction de la croûte due probablement à la fuite de gaz de l'intérieur. Une partie des mers lunaires aurait pris naissance à cette époque. L'apparition des cratères en général, devrait avoir lieu pendant la Téléosélène.

Selon le schéma précédent, l'état actuel de la surface lunaire, du point de vue morphologique, devrait être dû à l'action de forces endogéniques et exogéniques, agissant indépendamment ou successivement. L'observation montre que les impacts météoriques transforment des formations d'origine interne et que l'action interne transforme des formations d'origine météorique. A ces deux procédés d'élaboration de la surface lunaire, il faut ajouter l'érosion provoquée par le bombardement des rayons cosmiques, du vent solaire et surtout des micrométéorites et les excavations du matériau qui en résulte. Le matériau érodé pourrait se comporter comme une pondre qui, ayant attent les régions de déposition, pourrait se lithifier par soudage dans le vide, où les propriétés cohésives des substances permettraient de former des couches poreuses recouvrant le sol.

En ce qui concerne le tectonisme lunaire, l'étude du modèle de la fracture du sol, de la forme, de l'orientation et de la distribution des accidents entourant les grandes formations ou délimitant les mers, constitue un élément fondamental de la recherche géologique lunaire.

La structure des rainures et des crevasses et leur orientation par rapport aux formation voisines pourraient probablement être expliquées par un procédé endogénique et magmatique et par un effet des tensions latérales. Les rides observées dans les mers, pourraient être dues à des effets de compression. L'examen de certaines régions lunaires, surtout des mers et des grands cratères ainsi que la variation de leur caractère colorimétrique, indiqueraient l'existence de coulées de lave. Ce phénomène est particulièrement évident dans le système des rainures dont le profil montre diverses étapes de la coulée de la lave à travers des fractures du sol sous l'action des mouvements, effets de tension ou de compression du sol.

L'existence de dômes pourrait être justifiée par l'arrêt de gaz en dessous d'une croûte de lave, ou par la pression hydrostatique exercée par une lave fluide contre une croûte de lave sus-jacente.

3. Les résultats qui découlent déjà des expériences Ranger VII, VIII et IX, sont les suivants:

La structure des deux mers Mare Cognitum et Mare Tranquillitatis et d'une partie de l'arène d'Alphonsus est presque la même.

Les sillons observés dans les mers, sont des soulèvements du sol du même albedo et de la même couleur que la mer. L'effet des forces endogéniques y est évident. Leur structure fine consiste en des accidents en relief d'une longueur jusqu'à 1 kilomètre, d'une largeur jusqu'à 50 mètres et d'une hauteur jusqu'à 10 mètres au dessus du sillon.

Des montagnes telles que le pic central d'Alphonsus, pourraient être formées par des épanchements volcaniques ou constituées de cônes de cendres volcaniques. Leur blancheur éclatante pourrait être expliquée par un dépôt d'une substance qui blanchit vivement au cours des éruptions. Dans des cas analogues sur la Terre, la substance trouvée était le gypse et un composé d'oxyde de calcium.

Le nombre de cratères météoriques primaires dans les mers augmente dans un facteur quatre quand le diamètre diminue de moitié. Ce rapport persiste à peu près dans le cas des cratères météoriques secondaires qui se groupent autor des cratères primaires.

L'existence de cratères à effondrement dans les mers et les arènes d'Alphonsus et de Ptolemaeus est évidente, montrant une origine volcanique.

Les auréoles sombres entourant certains cratères semblent être constituées de cendres volcaniques.

On constate une forte érosion de la surface par bombardement micro- et macro-météorique qui, selon les estimations, pourrait s'étendre jusqu'à une profondeur d'environ un mètre.

L'action des gaz internes échappés à travers la surface, semble avoir joué un certain rôle dans la formation des cratères de grandes dimensions.

Les mers et l'arène d'Alphonsus pourraient contenir des matériaux fragmentés, comme aussi des crevasses à une grande profondeur au-dessous de la surface.

L'aspect des petites formations de quelques mètres de diamètre montre une surface tourmentée parcourue par des failles. Cet aspect n'est pas en faveur de l'existence d'un dépôt de poussière important recouvrant ces formations.

On n'observe pas de dépressions sur les sillons des mers, les cavités relatives ayant probablement été remplies de magma.

Des formations du type Karst sont souvent observées dans les mers. Dans ce cas, le sol présente de légères ondulations analogues à celles produites par le vent sur une couche de neige. Cette structure pourrait être attribuée en partie à des petits affaissements du sol selon un mécanisme de refroidissement ou de drainage du magma.

Dans les systèmes de rayonnements de Tycho et de Copernicus, on observe deux sortes de craterlets secondaires; les uns, à remparts sombres sont répartis presque régulièrement autour d'un cratère principal à des distances jusqu'à trois diamètres de ce dernier; les autres, à remparts clairs, ne présentent pas une répartition régulière en azimuth; on pourrait leur attribuer une origine résultant d'impacts cométaires.

Il y a une fréquence remarquable de cratères secondaires et une très faibel quantité de rochers. Ceci pourrait s'expliquer par la faible cohésion de la couche superficielle du sol, dans laquelle les rochers éjectés seraient ensevelis en créant des cratères secondaires.

Il y a un grand nombre de craterlets dans l'arène d'Alphonsus, ce qui n'est pas le cas pour ses remparts. On pourrait supposer une nature différente pour les matériaux constituant l'arène et le rempart.

Les rainures d'Alphonsus et la plupart de ses formations négatives proviennent d'effondrements. L'épine centrale de l'arène ainsi que les formations linéaires et les collines qu'on y observe, doivent être des produits d'activité endogénique. La persistance d'un lac de lave à la partie nord du rempart de ce cratère, indique une mobilité limitée du magma souterrain.

Des formations provenant d'éboulement sont rarement observées sur les remparts des cratères.

Les bordures des rainures sont arrondies probablement par érosion ou autres causes ayant rapport avec les étapes de solidification des matériaux de la surface.

b) Etat probable de nos connaissances sur la Lune lors de l'installation du LIL

Ces informations générales sur les propriétés physiques du sol lunaire seront complétées avant l'installation du LIL par l'application de nouvelles techniques d'observation détaillée depuis la Terre et par les programmes du type Luna, Ranger, Surveyor et Orbiter, Apollo et autres, qui pourront couvrir aussi l'hémisphère invisible de la Lune.

Les mesures Infra-Rouge auront fait connaître en grand détail les anomalies thermiques de la surface.

A l'aide de spectromètres appropriés on connaîtra l'intensité du rayonnement gamma qui caractérise le potassium-40. Cet élément permet l'étude de la différenciation magmatique qui a eu lieu sur la Lune.

Des mesures avec des spectrographes infra-rouge permettront des précisions sur les caractères minéralogiques des régions de la surface et les types des rochers.

Des mesures radar apporteront des nouvelles données sur l'épaisseur de la couche poudreuse recouvrant la surface et sa consistance.

Des mesures de la variation du gradient de gravitation permettront l'étude de la structure des couches au-dessous de la surface.

Des microscopes pétrographiques contrôlés depuis la Terre et des équipements spéciaux fourniront des mesures de la densité, de la conductivité thermique, des vitesses de propagation acoustiques, etc... Des échantillons de la surface du sol lunaire seront probablement déjà parvenus dans nos laboratoires terrestres.

Les cartes géologiques de certaines régions lunaires préparées actuellement par le United States Geological Survey couvriront probablement la surface entière de l'hémisphère visible et seront enrichies des données nouvelles recueillies à la suite de la réalisation des programmes précités, et ceux qui auront pour objet l'hémisphère invisible de la Lune.

Un tel travail préparatoire, achevé avant la création sur la Lune du LIL, permettra aux spécialistes de faire le choix des sites où seront installées les stations de recherche géologique lunaire.

c) Etudes géologiques à partir du LIL

L'aspect géologique lunaire ainsi établi servira de base à la recherche géologique directe après l'installation sur la Lune du LIL Celle-ci pourra être effectuée par des stations permanentes ou semi-permanentes reliées par des véhicules équipés d'instruments, et pourrait porter sur les objets suivants:

1. Le relevé de cartes morphologiques de la surface lunaire, à très haute résolution:

L'origine et l'évolution des formations de la surface d'après leurs caractères morphologiques pourront s'en trouver éclaircies.

2. L'étude de la rugosité du sol, et de la couche de poussière déposée sur les accidents lunaires, en différents endroits et selon leur structure; le cycle de production de la poussière par procédé de pulvérisation et la dilatation des magmas dans le vide pourront être précisés. L'âge relatif et absolu et la constitution chimique pourront être étudies. Le comportement du sol au cours de la lunaison, pendant laquelle la température varie entre $+125°$ et $-150°$ C, pourra être déterminé.

3. La variation des propriétés physiques des rochers en fonction de la profondeur. Ces études seront effectuées sur des échantillons obtenus à l'aide de forages, et pourront porter sur la radioactivité, la conductibilité thermique, la vitesse de propagation du son, la consistance, la mesure du flux de chaleur non solaire à la surface et en profondeur, la composition chimique en profondeur et sous la couche de poussière, etc. Ces forages seront effectués dans différents types de terrains, tels que des fonds de cratères, remparts, pitons centraux, mers, etc...

4. La recherche d'émanation de gaz de l'intérieur, leur localisation, l'étude de leur constitution chimique et abondance, et de leur origine.

La connaissance de la constitution chimique du sol lunaire, de l'abondance et de l'importance des éléments constituants ainsi que l'existence éventuelle de sources d'énergie telles que les émissions gazeuses, pourraient entraîner une exploration exhaustive de notre satellite en vue de son exploitation pour satisfaire les besoins des stations qui y seront installées.

Mais en outre, la recherche géologique, géochimique et géophysique sur la Lune présente un intérêt scientifique de première importance. La Lune, restée depuis son origine privée d'atmosphère appréciable, de variations climatiques à grande échelle et de sédimentation, se prête admirablement à l'étude de l'évolution des croûtes planétaires en remontant leur histoire jusqu'à un état primitif le plus reculé. La conquête de la Lune permettra aux astronomes d'aborder à partir de bases complètement nouvelles, le problème cosmologique de l'origine et de l'évolution du Système Solaire.

Références

J. VAN DIGGELEN, Photometric Properties of Lunar Crater Floors. Rech. Obs. Utrecht **14**, 1 (1959).

A. DOLLFUS, The Polarization of Moonlight, in: Physics and Astronomy of the Moon. New York: Academic Press, 1962.

A. DOLLFUS, Détermination de la nature du sol lunaire par la polarisation de la lumière. Colloquium, Physics of the Moon, R.A.S., London, 1965.

V. P. DZHAPIASHVILI and V. P. XANFOMALITI, Electronic Polarimetric Images of the Moon. Communication Comm., 16th Intern. Astr. Union Meeting, Hamburg, 1964.

G. FIELDER, Structure of the Moon's Surface. London: Pergamon Press, 1961.

G. FIELDER, Lunar Geology. London: Lutterwork Press, 1965.

J. E. GEAKE, Laboratory Investigations of Meteorite Luminescence. Colloquium, Physics of the Moon, R.A.S., London, 1965.

T. GOLD, Proton and Cosmic Ray Bombardment. Colloquium, Physics of the Moon, R.A.S., London, 1965.

T. GOLD, The Nature of the Moon's Surface as Derived from Physical Measurements. Colloquium, Physics of the Moon, R.A.S., London, 1965.

J. GREEN, The Geosciences Applied to Lunar Exploration, in: The Moon, Z. KOPAL and Z. K. MIKHAILOV, eds., p. 169. New York: Academic Press, 1962.

H. C. INGRAO and D. MENZEL, Lunar Research at Harvard College Observatory. Communication Comm., 16th Intern. Astr. Union Meeting, Hamburg, 1964.

G. P. KUIPER, The Lunar Surface and its Origin. Colloquium, Physics of the Moon, R.A.S., London, 1965.

G. P. KUIPER, Lunar Results from Rangers 7 to 9. Sky and Telescope, Special Supplement, p. 293, May 1965.

P. LOWMAY and D. BEATTIE, Manned Lunar Scientific Operations, American Astronautical Soc., Annual Meeting, May 4—7, 1964.

A. V. MARKOV, Nature of the Moon's Surface as Derived from Physical Measurements, Colloquium, Physics of the Moon, R.A.S., London, 1965.

B. C. MURRAY, Thermal Radiation during the Lunar Night Time. Colloquium, Physics of the Moon, R.A.S., London, 1965.

N.A.S.A., Ranger VII, VIII and IX Photographs of the Moon. Colloquium, Physics of the Moon, R.A.S., London, 1965.

G. PETTENGILL, Radar Observations, Colloquium, Physics of the Moon, R.A.S., London, 1965.

L. B. RONCA, An Introduction to the Geology of the Moon. Air Force Cambridge Research Laboratories, Special Report, N° 23, 1965.

S. K. RUMCORN, D. C. TOZER and J. WILSON, Electrical Conductivity of the Moon's Interior, etc ... Colloquium, Physics of the Moon, R.A.S., London, 1965.

A. SALOMONOVICH, Radiometric Observations of the Moon at 4 mm to 10 cm. Colloquium, Physics of the Moon, R.A.S., London, 1965.

R. W. SHORTILLE and F. M. SAARI, Radiometric and Photometric Mapping of the Moon through a Lunation. Communication Comm., 16th Intern. Astr. Union Meeting, Hamburg, 1964.

H. C. UREY, Study of the Ranger Pictures of the Moon. Colloquium, Physics of the Moon, R.A.S., London, 1965.

U.S. Geological Survey, Astrogeologic Studies, Annual Progress Reports, August 25, 1961 to July 1, 1964.

Sur un programme de sélénophysique pour un Laboratoire Lunaire International

Par

Georges Jobert[1]

Résumé — Abstract — Резюме

Sur un programme de sélénophysique pour un Laboratoire Lunaire International.
On suppose que toutes les analyses difficiles de matériaux lunaires pourront être faites
dans des laboratoires terrestres. L'activité des visiteurs devrait donc être partagée
entre l'acquisition d'échantillons et celle de données impossibles à acquérir à distance
ou automatiquement, ou la préparation d'expériences complexes.

Probablement, comme sur Terre, la méthode la plus efficace pour déterminer la
structure de la Lune sera la séismologie. Des essais antérieurs auront pu donner une
idée des signaux séismiques existants. Si leur fréquence et leur niveau sont suffisants,
il conviendra d'installer convenablement des stations fonctionnant ensuite de façon
autonome. Dans le cas contraire les expériences séismiques par explosion permettront
au moins l'étude des couches superficielles.

La détermination, à l'aide de gravimètres et d'inclinomètres, de la réponse de la
Lune aux forces de marée solaire et terrestre, peut permettre un choix entre différents
modèles. Mais pour un bon fonctionnement de ces appareils on devra sans aucun doute
les installer dans des caves assez profondes. Les grandes variations de la température
superficielle rendront également difficiles les mesures du flux thermique dont la détermination est de la plus grande importance.

Les techniques classiques de prospection géophysique devront bien entendu être
réexaminées, compte tenu de la situation particulière. L'absence probable d'un champ
magnétique planétaire n'enlève nullement leur intérêt aux mesures magnétiques
(magnétisme fossile éventuel, magnétisme des météorites . . .). En l'absence de magnétosphère propre les variations magnétiques en surface risquent d'être très faibles et il
est peu probable que la méthode magnéto-tellurique soit utilisable. Par contre, les
méthodes électriques semblent pouvoir être appliquées, dans des conditions peut-être
analogues à celles rencontrées dans les déserts terrestres. La prospection gravimétrique,
peut-être moins intéressante dans les premières phases d'occupation, sera sûrement
utilisée, vu sa simplicité, dans toute campagne d'exploration.

On a Program of Selenophysics for a LIL. It is assumed that all difficult analyses
of lunar materials can be made in terrestrial laboratories. The visitors' activities should
therefore be divided between securing samples and collecting data which are impossible
to obtain at a distance or automatically, or preparing complex experiments.

No doubt, like on the Earth, the most effective method of determining the Lunar
structure will be seismology. Earlier tests may have provided some idea of the existing
seismic signals. If their frequency and intensity are sufficient, it will be necessary to
install stations which will then function automatically. If not, seismic experiments
by explosion will make it possible to study the surface layers.

Determining the response of the Moon to solar and terrestrial tidal forces by

---

[1] Institut de Physique du Globe de Paris, France.

means of gravimeters and inclinometers may make it possible to choose between different models. But for these apparatus to function well, there is no doubt that they will have to be installed in quite deep cellars. The large variations in the surface temperature will also make it difficult to measure the thermic flux, the determination of which is of the greatest importance.

The classical techniques of geophysical prospection will, of course, have to be re-examined in consideration of this special situation. The probable absence of a planetary magnetic field in no way lessens the interest of magnetic measurements (possible fossile magnetism, magnetism of meteorites, etc.). In the absence of an actual magnetosphere, the magnetic variations at the surface may be very slight, and it is not very likely that the magnetotelluric method can be used. On the other hand, electric methods seem to be applicable in conditions which may be analogous to those encountered in terrestrial deserts. Gravimetric prospection, which is perhaps less interesting in the first stages of occupation, will certainly be used, because of its simplicity, in any exploration mission.

Программа селенофизических исследований для МЛЛ. Можно предположить, что все сложные анализы лунных материалов могут быть выполнены лабораториями, находящимися на Земле. Поэтому деятельность членов экспедиции на Луне должна ограничиваться сбором образцов и данных, которые невозможно получить на расстоянии или автоматически, или подготовкой к выполнению сложных экспериментов.

Вероятно, как и на Земле, наиболее эффективным методом определения структуры Луны будет сейсмология. Опыты, которые будут проведены предварительно, смогут дать представление о существующих сейсмических сигналах. Если они будут иметь нужную частоту и нужный уровень, то потребуется должным образом установить станции, которые в дальнейшем будут работать автономно. В противном случае сейсмические эксперименты с помощью взрывов позволят изучить, по крайней мере, поверхностные слои.

Определение с помощью гравиметров и уклономеров реакции Луны на солнечные и земные приливы может позволить остановить выбор на определенных моделях. Но для того, чтобы эти приборы хорошо работали, их потребуется установить в довольно глубоких колодцах. Большие колебания поверхностной температуры также затруднят измерения термического потока, определение которого имеет очень большое значение.

Классические методы геофизических исследований должны быть, разумеется, пересмотрены с учетом специфики условий. Вероятно, отсутствие магнитного поля у Луны ни в коей мере не снижает интерес магнитных измерений (возможный остаточный магнетизм, магнетизм метеоритов и т.д.). Ввиду отсутствия собственного магнитного поля, вариация магнитных сил на Луне может оказаться чрезвычайно незначительной, в связи с чем применение магнитно-теллурических методов, вероятно, окажется невозможным. Напротив, электрические методы, вероятно, можно будет использовать подобно тому, как они используются в условиях пустынь на Земле. Гравиметрические исследования, вероятно менее интересные на первой фазе работы экспедиции на Луне, безусловно, ввиду их простоты, будут проводиться в ходе любых исследовательских кампаний.

Tout programme de recherche pour un laboratoire sur la Lune doit être conçu compte-tenu de la progression technologique qui a conduit à son installation: stations de mesures automatiques, raids de véhicules télécommandés, satellites circumplanétaires habités, débarquements de courte durée — et des étapes ultérieures que l'on peut envisager: installation permanente et exploitation des ressources locales. Bien entendu les renseignements obtenus dans les premières étapes serviront de base pour la préparation du programme définitif et amèneront éventuellement à le modifier sur certains points.

De toute façon les moyens limités correspondant à une installation temporaire ne permettront pas une activité aussi étendue que celle d'un laboratoire terrestre. L'activité des physiciens consistera à installer un réseau de stations de mesure, en effectuant pendant les raids le maximum de mesures physiques, et à monter des expériences difficiles à mettre en place dans une station télécommandée ou automatique.

En principe l'activité d'un laboratoire de sélénophysique devrait se partager, comme celle des laboratoires de géophysique, entre l'étude des parties externes de la planète et celle de ses parties internes. L'absence très probable d'atmosphère et de magnétosphère restreint toutefois considérablement le champ des études externes, ou plutôt conduit à les rattacher étroitement à l'astrophysique (étude du champ magnétique interplanétaire, du vent solaire . . .). Toutefois un programme limité de détection des émanations et des gaz issus du sol, et d'enregistrement des variations magnétiques rapides devra être prévu pour les premières stations automatiques; les résultats obtenus pourront conduire à un programme spécifique pour le LIL. Nous nous bornerons ici à l'étude des propriétés internes de la Lune.

C'est bien entendu la séismologie qui occupera la plus grande place dans ce programme. C'est à elle que l'on doit la plus grande part de nos connaissances actuelles sur l'intérieur de la Terre aussi bien pour ses couches les plus superficielles que pour les plus profondes. On peut penser qu'elle jouera le même rôle dans le cas de la Lune. L'installation d'une station séismographique automatique — réalisable dans un avenir assez proche — fournira d'abord une réponse à la question fondamentale: se produit-il assez fréquemment sur la Lune des signaux séismiques ?

On peut envisager trois sources possibles de tels signaux. L'activité tectonique est considérée en général comme la source la plus probable. Des calculs sur l'évolution thermique de la Lune, basés sur des hypothèses concernant sa composition et sa richesse en matériaux radioactifs, conduisent en effet à des débits d'énergie considérables. On assimile, dans ces calculs, le matériau lunaire soit au manteau et à la croûte terrestres, pour lesquels on ignore malheureusement la distribution exacte des éléments radioactifs en profondeur, soit à certains types de météorites, les chondrites en général. Dans le cas d'une planète homogène, de composition chondritique, l'énergie thermique libérée par les désintégrations radioactives, a été évaluée à $4 \cdot 10^{24}$ ergs/an. Cette hypothèse a été discutée, car elle conduirait à des températures internes élevées, voisines du point de fusion des silicates ou le dépassant. Si l'on admet une richesse en produits radioactifs moindre, ou une différenciation ayant amené en surface la plus grande partie de ces derniers, on aura bien entendu des débits plus faibles. Rappelons que sur Terre l'énergie libérée par les séismes est de l'ordre de 10^{25} ergs/an.

Il faut toutefois remarquer que la liaison probable existant entre tremblements de terre et régime thermodynamique est loin d'être comprise et que, d'autre part, les structures géologiques liées sur Terre aux zones séismiques semblent faire défaut à la Lune. Il est donc possible que si les séismes y existent, ils soient d'un type différent de ceux auxquels nous sommes habitués. Le caractère essentiel du phénomène en définitive est son début brusque, qui permet la propagation d'un "front avant" pour les ondes séismiques. Ce qui suit modifie la forme des enregistrements, mais n'empêche pas leur exploitation. Il est possible que ce caractère fasse défaut aux séismes lunaires, ce qui rendrait considérablement plus difficile la tâche des séismologues.

D'éventuels phénomènes volcaniques pourraient aussi être la source de signaux séismiques. On sait toutefois que sur Terre les séismes d'origine volcanique sont de faible magnitude et qu'ils fournissent des enregistrements difficilement utilisables pour l'étude des couches profondes, les phases y manquant de netteté.

Enfin la chute de météorites de masse suffisante pourrait provoquer des ondes séismiques décelables à distance, comme l'a montré celle de la météorite Tunguska, le 30 Juin 1908, pour laquelle l'énergie libérée a été évaluée entre 10^{21} et 10^{23} ergs.

On voit que sur Terre le phénomène est rare. L'absence d'atmosphère sur la Lune permet sans doute l'arrivée plus fréquente de météorites intactes. On a évalué le nombre d'impacts détectables en une année par un appareil dont le seuil de sensibilité se situerait vers 1 mμ en utilisant des estimations du nombre de chutes. Ce nombre varie entre deux douzaines et moins d'une unité, suivant le pouvoir absorbant que l'on attribue au matériau qui constitue l'intérieur de la Lune. Les signaux fournis par ces chutes seraient aussi utilisables que ceux de véritables séismes, mais ils devraient en être distingués pour l'étude de la séismicité de la Lune, et on rencontrerait dans ce problème les mêmes difficultés que sur Terre dans la discrimination entre séismes et explosions nucléaires.

Une autre information sera fournie en tous cas par la station automatique : le niveau du bruit de fond et sa distribution spectrale, utile pour filtrer une éventuelle agitation microséismique. L'absence de masses mobiles (océans, atmosphère) en surface permet d'espérer un niveau beaucoup plus bas que sur Terre. On pourrait penser que les déformations des roches, causées par les grands changements de température, provoquent des ruptures, sources de bruit. Mais sur la plus grande partie de la surface, les roches ont été exposées à ces variations depuis des millions d'années et ont dû atteindre un état d'équilibre, soit qu'elles se soient finement divisées, soit qu'elles soient recouvertes d'une couche pulvérulente d'origine différente qui les mette à l'abri de ces variations. Dans le cas où cette protection serait insuffisante, l'étude de la variation du bruit au cours du temps pourrait présenter quelque intérêt, les fronts de changements brusque de température se déplaçant assez rapidement à la surface.

Admettons d'abord que des signaux suffisants existent, quelle que soit leur origine. Leur exploitation doit permettre d'une part la détermination des coordonnées de la source pour définir les zones séismiques de la Lune ou le point de chute d'une météorite, d'autre part celle des vitesses internes. Avec une seule station on ne dispose pas d'assez d'information et l'on doit faire appel à des résultats théoriques. A partir de modèles de répartition de la densité, de l'incompressibilité et du module de Poisson, on détermine les courbes de propagation des ondes longitudinales P et transversales S. De la différence observée entre les temps d'arrivée des ondes P et S on déduit ainsi la distance épicentrale. L'azimut de l'épicentre et la profondeur du foyer sont beaucoup plus difficiles à évaluer.

Pour déterminer expérimentalement la répartition des vitesses internes, il est d'abord nécessaire d'obtenir une bonne précision dans le tracé des courbes de propagation. L'amélioration de ce tracé s'effectue par approximations successives et nécessite un réseau de stations convenablement installées, permettant une bonne trilatération. Remarquons en passant que l'existence même de courbes valables pour l'ensemble de la planète, quelles que soient les positions relatives de la station et de l'épicentre, n'est possible que si sa structure interne possède une symétrie sphérique. Selon certaines vues la Lune serait un objet à structure beaucoup plus complexe. Dans ce cas les problèmes d'inversion seraient très difficiles à résoudre.

D'autres méthodes, basées sur l'étude des ondes superficielles, pourront bien entendu être utilisées avant qu'un réseau soit installé. Par exemple l'observation d'ondes ayant fait plusieurs fois le tour de la Lune pourrait fournir des renseignements sur leur vitesse de groupe et sur les propriétés anélastiques de l'intérieur. Les courbes de dispersion obtenues permettraient de choisir entre différents modèles théoriques.

En conclusion, si des essais préliminaires ont mis en évidence des séismes d'importance suffisante, la première mission du laboratoire sera d'installer plusieurs stations bien réparties, aussi complètes que possible. Les signaux obtenus pourront être enregistrés à la station principale et réémis de là vers la Terre. Un traitement rapide des données permettra d'effectuer des reconnaissances sur les lieux où un phénomène intéressant se sera produit.

Dans le cas où les signaux seraient trop rares ou trop faibles devant le bruit, on devrait tenter d'utiliser les variations de ce dernier, et de toute façon appliquer les méthodes de prospection classiques. L'inconvénient de ces dernières est l'importance des moyens à mettre en œuvre. On pourrait, dans une première étape, se borner à étudier la structure des couches superficielles au voisinage de LIL, soit en effectuant des profils à l'aide de véhicules, soit en tirant des charges à explosion télécommandée depuis la station. Des profils de plusieurs hectomètres pourraient même se faire automatiquement. En l'absence totale de séismes on serait amené à effectuer des tirs très importants une fois un réseau assez dense de stations établi sur la Lune.

On sait qu'avant le développement de la séismologie, on avait pensé tirer parti des marées du Globe terrestre excitées par les variations d'attraction de la Lune et du Soleil. Les résultats de ces études, malgré leur développement considérable depuis l'A.G.I., sont encore de peu de poids devant ceux fournis par l'étude des séismogrammes. On peut au plus vérifier qu'ils sont compatibles avec les modèles des séismologues. Dans le cas de la Lune on sera, au moins initialement, heureux de disposer du maximum d'information possible. Que peut donner l'étude des variations de la pesanteur lunaire, en direction et en intensité ?

On a évalué récemment les amplitudes théoriques de ces variations pour un modèle de Lune indéformable et les déviations que produirait la déformation pour différents modèles. La conclusion est assez peu encourageante: l'amplification des variations de l'intensité de la pesanteur pour la déformation serait de l'ordre de 1% pour tous les modèles envisagés, sauf pour ceux dans lesquels un noyau fluide occuperait les $^3/_4$ de la planète, pour lesquels la modification atteindrait 5%. La réduction correspondante des variations de la direction de la pesanteur serait plus importante, de l'ordre de 5% en général, mais atteignant 14% pour un gros noyau.

L'amplitude totale des variations — dues essentiellement aux variations de la distance Terre-Lune et à la libration au cours de la lunaison — est assez grande: 6 µrad et 1 mgal (au lieu de 0,2 µrad et 0,3 mgal pour la Terre). La faible valeur de la pesanteur lunaire a aussi pour effet de faciliter en principe les conditions d'opération. Par exemple le ressort d'un gravimètre terrestre devra supporter une masse six fois plus importante que sur la Terre et un changement donné de pesanteur correspondra donc à une variation de tension six fois plus importante, alors que la dérive liée aux propriétés du ressort ne sera pas modifiée. Mais malgré tout, l'élimination de cette dernière reste le problème essentiel. En effet les variations cherchées ne sont plus diurnes et semi-diurnes, comme sur Terre, mais mensuelles et semi-mensuelles. Les difficultés rencontrées actuellement pour mettre en évidence les marées de grande période tiennent surtout aux irrégularités de la dérive des instruments. Il sera donc nécessaire de choisir des instruments ayant fait la preuve d'une très grande stabilité et de les installer dans des stations particulièrement bien protégées. Pendant la lunaison la température de la surface subit des changements de l'ordre de 300° C. L'évaluation du signal thermique périodique ainsi engendré risque d'être très difficile et son élimination complète est improbable.

La connaissance de la répartition dans l'espace du champ de pesanteur lunaire sera également très utile car ses caractéristiques dépendent à la fois de la figure de la Lune et de la répartition des masses internes. La détermination précise des caractéristiques orbitales de satellites de la Lune semble difficile à faire depuis des stations de repérage sur la Terre. Un réseau de stations sur la Lune permettrait une bien meilleure précision et par conséquent une définition plus fine du champ de pesanteur lunaire.

Passons maintenant aux recherches à effectuer sur le terrain.

Un des paramètres les plus intéressants à mesurer sur la Lune est le flux de chaleur d'origine interne. Contrairement à ce qui se passe pour les continents de notre planète à cause des glaciations quaternaires, la température moyenne superficielle en un point donné de la Lune a dû peu varier — sauf évidemment dans une région volcanique. Il suffit donc de faire la mesure à une profondeur suffisante pour que les variations mensuelle et annuelle — et éventuellement undécennale — y soient complètement annulées. Cette profondeur dépend de la diffusivité du sol, facteur que l'on devra évaluer le plus tôt possible. Les mesures du rayonnement thermique de la Lune conduisent à estimer à 10^{-5} ou 10^{-6} cal/cm s °C la conductivité thermique des couches tout à fait superficielles. En prenant 1,5 g/cm³ pour leur densité, 0,2 cal/g °C pour leur chaleur spécifique, on obtient 10^{-5} cm²/s pour la diffusivité. La profondeur où la variation annuelle est réduite à $\exp(-2\pi)$, soit 0,002, fois sa valeur superficielle, est donc de l'ordre de 70 cm $\left(l = \left[\dfrac{4\pi\,kP}{\varrho\,c} \right]^{1/2} \right)$.

La perturbation mensuelle du flux est en surface d'environ 130 μcal/cm² s, ou 500 ergs, soit 500 à 50 fois le flux d'origine interne probable. Mais la profondeur de pénétration de cette onde est $\sqrt{12}$ fois plus petite que celle de l'onde annuelle, et à 70 cm son amplitude est réduite d'environ 10^{-9} fois. La très faible conductivité thermique des couches superficielles permettra donc de faire les mesures sans difficulté. Toutefois la richesse radioactive des poussières superficielles peut différer de celle des couches plus profondes, et des corrections devront être faites après détermination du relief caché.

Nous n'entrerons pas dans le détail des théories concernant l'origine du flux sélénothermique (radioactivité, frottement interne...). Remarquons seulement que la mesure du flux permettra peut-être de rejeter certains modèles proposés. Plus probablement la plupart des théories s'accommoderont aisément des valeurs obtenues.

Les recherches sur le magnétisme seront effectuées dans des conditions radicalement différentes de celles connues sur Terre. Il apparaît en effet comme extrêmement probable que la Lune ne possède pas un champ magnétique propre — ce qui est compréhensible si un tel champ est dû à l'action de courants à l'intérieur d'un noyau fluide conducteur. Néanmoins les matériaux du sol lunaire sont aimantables et sont plongés dans le champ magnétique interplanétaire. On sait encore peu de choses sur ce dernier, sinon qu'il est très faible (de l'ordre de $10\,\gamma$ ou 10^{-4} Oe). Une station magnétique automatique fournira des données indispensables sur ses variations en grandeur et en direction. Si le champ a un comportement très irrégulier (ce qui est possible vu sa liaison avec le vent solaire) on aura beaucoup de difficultés à utiliser ce paramètre physique. Si au contraire une certaine constance se manifeste, on peut envisager diverses études. Par exemple si les mers correspondent à des épanchements de lave, solidifiés, on pourra étudier leur aimantation thermorémanente, comparer les champs existant à l'époque de leur solidification. Ces études ne paraissent pas devoir conduire à une construction aussi intéressante que celle que permet le champ terrestre dans le domaine de l'archéo- et du paléomagnétisme. Mais de toute façon la détermination des propriétés magnétiques intrinsèques des roches (susceptibilité, caractéristiques des courbes d'aimantation...) permettra d'utiles classements.

Bien entendu tous les raids effectués à partir du LIL seront l'occasion de mesures systématiques de magnétisme, radioactivité et gravité. Les profils obtenus permettront sans doute de choisir entre les deux explications proposées pour l'origine des reliefs lunaires — volcanisme ou chute de météorites. Si l'on peut effectuer de bonnes déterminations géodésiques, un autre problème pourra être abordé: la surface des mers est-elle une surface de niveau du champ de pesanteur actuel? La mise en évidence d'un écart notable entre ces surfaces aurait des conséquences théoriques considérables.

Les autres méthodes de prospection devront être adaptées au cas particulier que constitue la surface lunaire. L'étude des couches les plus superficielles intéressant les géologues, sera faite par sondages de résistivité électrique, sondages neutron-neutron ou neutron-gamma (pour mettre en évidence d'éventuelles couches de glace ou des dépôts d'éléments à grande section efficace, comme le soufre, le chlore ou le bore), sondages gamma pour déterminer les densités, etc... Une technique permettant une plus grande profondeur d'investigation est le sondage magnétotellurique, qui n'est toutefois possible qu'en présence de phénomènes périodiques ou à constante de temps définie. Il est malheureusement peu probable que l'on observe de tels phénomènes sur la Lune en l'absence de magnétosphère. La réponse sera donnée par la station magnétique automatique.

En conclusion il semble que l'on puisse espérer déterminer la structure interne de la Lune — et en particulier savoir si elle possède à peu près une symétrie sphérique ou au contraire si elle est constituée de blocs hétérogènes — par l'établissement d'un réseau de stations séismiques, si des séismes naturels se produisent fréquemment. Simultanément la détermination précise des orbites de satellites lunaires permettrait de savoir s'il existe des anomalies régionales du champ de pesanteur. Les résultats qui seraient fournis par une station unique ne permettraient pas de donner une réponse satisfaisante aux nombreuses questions qui seront posées.

Bibliographie

M. CAPUTO, On the Shape, Gravity Field, and Strength of the Moon. J. Geophys. Res. 70, 3993 (1965).

J. GREEN, Geosciences Applied to Lunar Exploration, in: The Moon, KOPAL et MIKHAILOV, eds. New York: Academic Press, 1962.

J. C. HARRISON, An Analysis of Lunar Tides. J. Geophys. Res. 68, 4269 (1963).

G. J. F. MAC DONALD, On the Internal Constitution of the Inner Planets. J. Geophys. Res. 67, 2953 (1962).

F. PRESS, P. BUWALDA and M. NEUGEBAUER, A Lunar Seismic Experiment. J. Geophys. Res. 65, 3097 (1960).

W. M. SINTON, Temperatures of the Lunar Surface, in: Physics and Astronomy of the Moon, KOPAL, ed., Ch. 11, p. 407. New York: Academic Press, 1962.

G. H. SUTTON, N. S. NEIDELL and R. L. KOVACH, Theoretical Tides on a Rigid Spherical Moon. J. Geophys. Res. 68, 4261 (1963).

H. C. UREY, Origin and History of the Moon, in: Physics and Astronomy of the Moon, KOPAL, ed., Ch. 13, p. 481. New York: Academic Press, 1962.

Lunar Meteorological Observatory for Observations of the Earth

By

K. Ya. Kondratiev[1], V. L. Gaevsky[2], V. N. Konashenok[2] and A. I. Reshetnikov[2]

(With 5 Figures)

Abstract — Résumé — Резюме

Lunar Meteorological Observatory for Observations of the Earth. The principal advantages of carrying out meteorological and solar observations from the Moon are considered. These are absence of a lunar atmosphere, low gravitation for the construction of equipment and the possibility of continuous observation of portions of the Earth's surface and of the Sun. The disadvantages of great distance from the Earth and motions of the Moon are reviewed.

The possibility of studying the Earth's cloud cover distribution is discussed in terms of instrumental requirements and location of lunar observatories.

It is pointed out that one of the principal tasks of a lunar observatory will be to find the relationships of solar activity to terrestrial atmospheric processes. The present understanding of this relationships is reviewed. The possibility of investigating the heat balance of the Earth's surface-atmosphere system is also considered.

Although a lunar observatory cannot replace meteorological satellites, complex investigations can be undertaken utilizing both locations for coordinated measurements.

Finally, the prospects of making indirect meteorological soundings of the Earth's atmosphere from the Moon are analysed. These include determination of the vertical temperature profile, the upper cloud cover boundary, ozone concentration, study of the horizontal optical non-homogeneity of the atmosphere by laser technique, passive radio-location of cloud formations, etc.

The authors conclude that a meteorological observatory on the Moon can play a great role in improving weather and climatic forecasting on the Earth.

Observatoire météorologique lunaire par l'Observation de la Terre. On examine les principaux avantages que présentent les observations météorologiques et solaires à partir de la Lune. Ceux-ci sont notamment: l'absence d'atmosphère lunaire, la faible gravitation pour la construction des équipements et la possibilité d'observer de façon continue des parties de la surface terrestre ou solaire. On examine les inconvénients provoqués par la grande distance depuis la Terre et les mouvements de la Lune.

On discute, du point de vue des spécifications des instruments et de l'emplacement des observatoires lunaires, de la possibilité d'étudier la distribution des nuages recouvrant la Terre.

On souligne que l'une des tâches principales d'un observatoire lunaire sera de trouver les relations entre l'activité solaire et les phénomènes atmosphériques terrestres. On passe en revue les connaissances actuelles sur ces relations. On examine également la possibilité d'étudier l'équilibre thermique du système surface terrestre — atmosphère.

Bien qu'un observatoire lunaire ne puisse pas remplacer les satellites météorologi-

---

[1] Rector, University of Leningrad, U.S.S.R.
[2] University of Leningrad, U.S.S.R.

ques, des recherches complexes peuvent être entreprises en utilisant les deux emplacements (surface lunaire et orbite terrestre) pour des mesures coordonnés.

On analyse enfin les perspectives de sondage météorologiques indirects de l'atmosphère terrestre à partir de la Lune. Ceux-ci comprennent la détermination du profil vertical des températures, de la limite supérieure de la couche de nuages, de la concentration d'ozone, l'étude des inhomogénéités optiques horizontales de l'atmosphère à l'aide de lasers, la radiolocalisation passive de formations de nuages, etc.

Les auteurs concluent qu'un observatoire météorologiques peut jouer un grand rôle dans l'amélioration des prévisions climatiques et de température sur Terre.

Лунная метеорологическая обсерватория для наблюдений Земли. Рассмотрены основные преимущества наблюдений Земли и Солнца с поверхности Луны: отсутствие лунной атмосферы, малая величина силы тяжести (что важно для установки оборудования и возведения сооружений), возможность непрерывных наблюдений части земной поверхности и Солнца. Обсуждены трудности, обусловленные большим расстоянием Земли от Луны и особенностями движения Луны.

Проанализированы требования к аппаратуре и местонахождению лунной обсерватории в связи с решением задачи исследования распределения облачного покрова на Земле.

Отмечено, что одной из главных задач Лунной обсерватории будет изучение связи между солнечной активностью и процессами в земной атмосфере. Сделан краткий обзор современного состояния исследований в этой области. Рассмотрена также возможность изучения теплового баланса системы „земная поверхность-атмосфера".

Подчеркнуто, что лунную обсерваторию не следует рассматривать как замену метеорологических спутников: решение проблемы может быть найдено лишь на основе использования комплекса данных наблюдений.

Обсуждены перспективы косвенных метеорологических зондирований земной атмосферы с поверхности Луны, включающих определение вертикального профиля температуры, высоты верхней границы облаков, распределения концентрации озона, изучение горизонтальной оптической неоднородности атмосферы с помощью лазеров, пассивную радиолокацию облачных образований.

Авторы отмечают в заключение, что метеорологическая обсерватория на Луне может сыграть большую роль в усовершенствовании прогнозов погоды и климата на Земле.

The rapid development of cosmonautics and the wonderful achievements made in this field in a short space of time bring near the day when Man will be able to land on the Moon. Such an event will be of first order importance not only for the study of the Moon, but also for many-sided investigations in Astrophysics, Geophysics and Meteorology. The aim of the present paper is to consider the prospects of meteorological investigations from the Moon's surface.

1. Principal Advantages of Observations from the Moon's Surface

The principal advantage for astrophysical and meteorological investigations from the Moon is its lack of atmosphere. According to recent data the density of the lunar atmosphere, if there is one, cannot exceed 10^{-12} g/cm³ though how much less it is still remains unknown. With the best instrumental observations of the Moon from the Earth we can distinguish objects whose dimensions are not smaller than 250 m. This limit is determined not by the qualities of optical instruments, but rather by atmospheric disturbances. When observing the Earth from the Moon we shall see it through the same atmospheric layer, but it will be much closer to the target of observation than to the telescope, therefore, from the Moon one will be able to dinstinguish objects 4 to 5 times smaller than those seen on the Moon from the Earth.

As to observations of the Sun whose emission is the source of energy for all weather-forming processes on the Earth, we have really infinite possibilities; for when conducting observations from the Earth we can observe the Sun only through several atmospheric transparency windows "given" to us by nature. Thus in performing observations from the Earth's surface we are deprived of the opportunity of obtaining data on the ultraviolet part of the spectrum, particularly the far-ultraviolet part, as well as on X-rays and γ solar emission. Unfortunately, the scrappy data lately obtained from satellites and rockets do not give us a clear picture of short-wave solar emission.

Construction of the Moon meteorological station will be facilitated by the absence of wind loadings and by much smaller gravity on the Moon. Since the winds and moving dust capable of contaminating lens and mirror surfaces are absent on the Moon, the weight of all optical systems may be much lowered.

The speed of the rotation of the Moon around its axis is equal to that of its revolution around the Earth, for this reason the Earth is always faced by the same hemisphere of the Moon, which permits constant observations of the development of weather-forming processes over the areas of the globe unavailable for land observations.

The amount of solar energy on a unit of a lunar surface is actually tremendous. For instance, if the Sun is in zenith, each square kilometre of the lunar surface receives in a minute 2×10^{10} cal of heat or 5.0×10^7 kw per hour (note, for comparison, that the power of the Dnieper Power Station is 9×10^5 kw per hour). Thus, if special thermal and energy accumulation systems are constructed, it will be possible to supply with energy any powerful observatory, with an entire town attached to it, for the period of the lunar night.

While quantum or thermal radiation detectors working in vacuum have much higher sensitivity they can also ensure the recording of more subtle processes. The application of quantum radiation detectors (which require cooling by liquid nitrogen under usual terrestrial condition) on the unlit and cooled side of the Moon will permit considerable simplification of receiving systems.

As a majority of thermal radiation detectors operate best of all at temperatures of $30-40°$ C, which can be ensured by protecting the equipment with appropriate reflecting shields, they can be employed as well on the sunlit side of the Moon.

Because of the large curvature relative to the small lunar ball the irregularities of the lunar surface will disappear from the field of view of instruments. The Moon's horizon will seem flat and monotonous to an observer.

Difficulties awaiting us on the Moon from the view-point of similar measurements and observations from satellites are as follows: first, a very large distance from the Earth. As is known, the Moon revolves along an ellipse, 363,300 km in perigee and 405,500 km in apogee, the eccentricity being equal to $E=0.0549$. Accordingly, angular dimensions of the Earth's hemisphere will vary from $2°00'$ in perigee (without the atmosphere) to $1°50'$ in apogee.

Second, there are periodic longitudinal and latitudinal librations. Longitudinal librations are associated with the non-uniformity of the orbital movement of the Moon (its revolution around its axis being uniform) whereas, the latitudinal librations are determined by the slope of the lunar orbit to the ecliptic. Thus, when designing Earth observation systems one will have to provide them with devices for compensating such inequalities in the motion of the Moon.

As to the observations of the Sun they can be carried out by one station for more than 13.5 days. For the remaining time the station would be on the unlit side of the Moon. The way out may be found in using either two stations on opposite sides of the lunar ball or better, three — situated at an angular distance of 120° from one another.

The choice of the optimum number of lunar observatories is dependent on the possibility of conducting continuous observations of the motion of the Earth and the Sun in the lunar sky. As has been mentioned before, the general character of the Earth's motion relative to the lunar horizon is determined by latitudinal and longitudinal librations.

An observer at a fixed point on the lunar surface will see the Earth during the whole period of observation in the same section of the celestial sphere with the approximate dimensions of 20° in azimuth and 20° in altitude. The Earth's centre will move along a curve of Leisageaux shape. The position of the Earth relative to the horizon at a fixed moment is determined by ephemerises. Formulas for the Earth's ephemerises in horizontal coordinates have been derived by V. V. Schev-chenko [2]. With these formulas and data from the Astronomical Annual the Earth's coordinates in the lunar sky can be easily calculated for any moment of time.

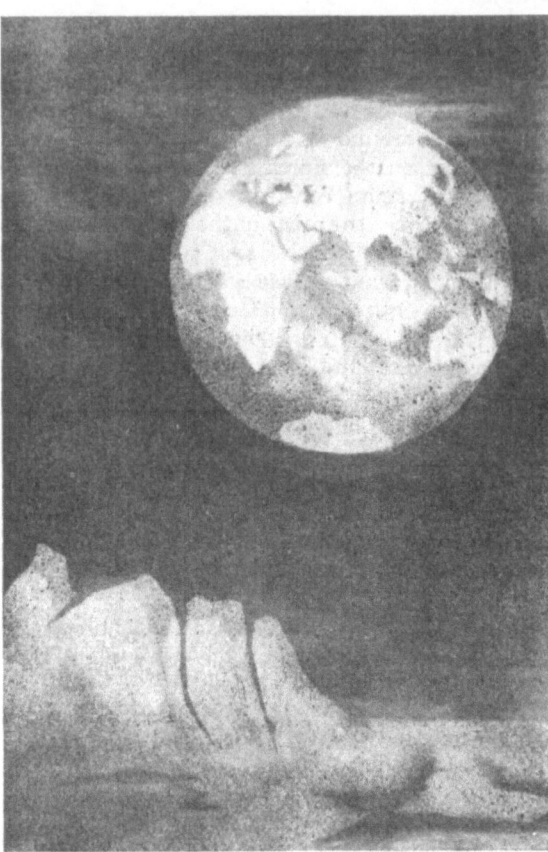

Fig. 1. Schematic image of the Earth as visually observed from the Moon

The highest position of the Earth over the lunar horizon will be seen at the point of selenographic coordinates $\lambda = 0, \beta = 0$ (if these values are ascribed to the central point of the lunar disk as seen from the Earth). During the motion along the lunar equator from zero meridian the Earth altitude over the horizon decreases. From the edges of the lunar disk the Earth will be seen nearly on the horizon.

The portion of the Earth's surface available for observations from the Moon at a given moment can be determined with the help of astronomical tables.

After examining the interposition of the Sun, the Earth and the Moon it becomes evident that the phases of the Earth in the lunar sky will be opposite to those of the Moon. Which portion of the visible lunar disk is sunlit is known from astronomical tables, which also give the values of K of the sunlit portion of the Moon diameter perpendicular to the terminator. Because of the additional Earth and Moon phases for each moment of time, the $1 - K$ sunlit portion of the Earth diameter perpendicular to terminator will be seen by an observer on the Moon.

Thus, from an observatory situated on the visible side of the Moon it is possible to carry out continuous observations of a certain portion of the Earth. Coordinates of the limit of the Earth's visible portion may be easily computed for any moment of time.

2. Possibilities of Visual and Instrumental Observations of the Planetary Distribution of Cloud Cover. Necessary Spatial Resolution

It is known that a man's eye can perceive an angular resolution equal to 1′ when the illumination is 1 lux. In case of a higher intensity of illumination the angular resolution of an eye increases up to 0.7′. So, an observer on the Moon not provided with any optical instrument will be able to distinguish cloud forma-

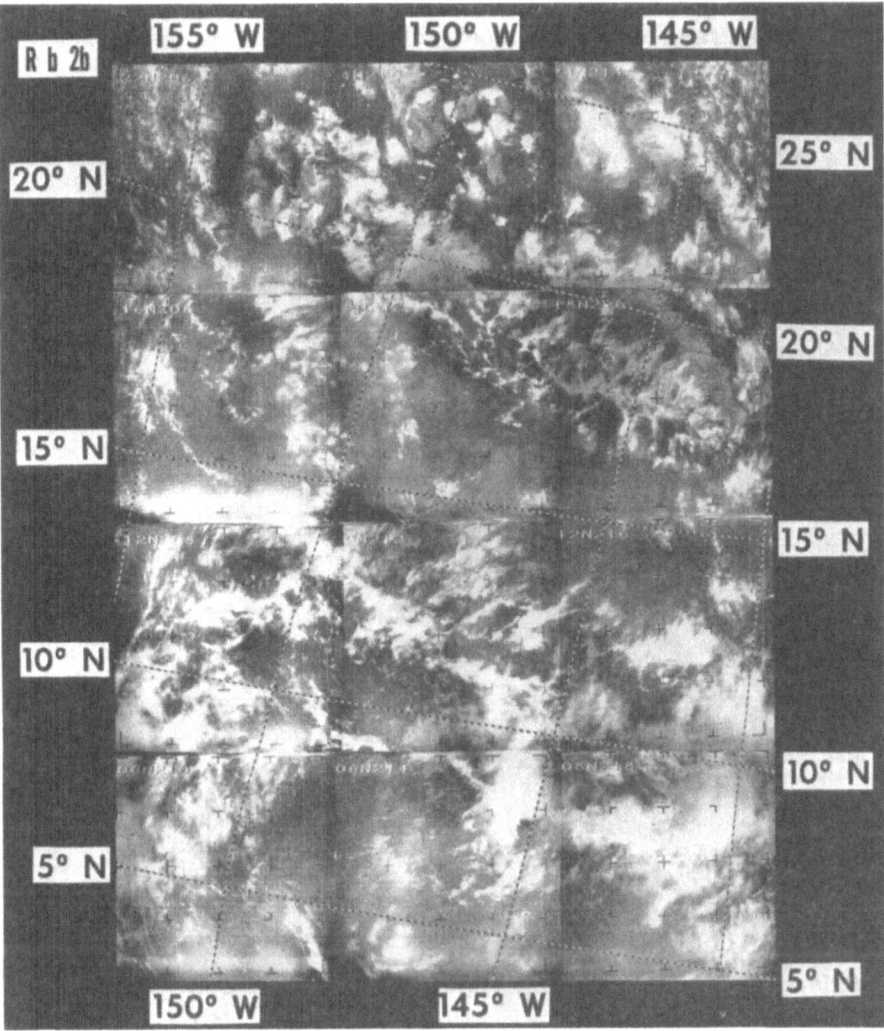

Fig. 2. Complex photograph obtained by T.V. camera with a high resolution from the "Nimbus" satellite

tions of about 100×100 km in area. As an example, let us examine Fig. 1 representing schematically the image of the Earth as observed visually from the Moon. The Earth's surface is distinctly seen. The Black and Caspian Seas and even Riga Bay are easily distinguished, and a cloud formation over the Indian Ocean is well seen. This shows that an observer will be able to detect easily, for instance, such things as spiral cloud bands associated with cyclonic storms from 800 to 1,500 km in diameter [1].

Using binoculars—the simplest optical instrument—an observer will be able to see the regions of a hurricane formation, to distinguish cellulare cloud systems with the cells of 60—80 km in diameter and even large cumulus.

By employing a telescope of simple design one will be able to get a resolution of several angular seconds. Thus an observer will be able to watch nearly all the peculiarities of the atmospheric circulation.

Since it is possible to attain the theoretical threshold of instrumental resolution of about 0.15″, and due to the lack of an atmosphere on the Moon, the use of powerful telescopes on the Moon would also be rather promising.

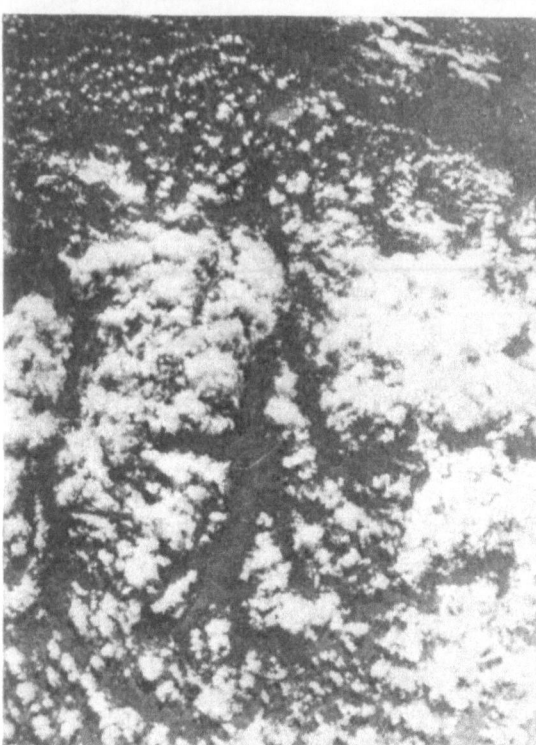

Fig. 3. Photograph taken by Soviet cosmonauts with a resolution of 1—1.5 km

Let us examine some of the pictures of the Earth taken from satellites, and interpret them from the point of view of observations from the Moon's surface.

Fig. 2 illustrates a complex photograph obtained by a AVCS T.V. camera, with a high resolving power, from the "Nimbus" satellite. For the altitude of a satellite relative to the Earth's surface of about 1,000 km, the resolution came out to be better than 0.9 km, and the viwed area of the Earth's surface, of approximately square shape, was equal to 630 × 630 km. The angle of view of the camera is about 35°. The same resolution when observing from the Moon is ensured by a telescope with an angle of view equal to 7′, and an angular resolution equal to 0.7″.

The picture in Fig. 3 taken by Soviet cosmonauts is reproduced from the booklet "Our Planet from Space" [3]. The photograph was taken from an altitude of 200 km with a camera whose angle of view was 41°; the field of view is, approximately, 150 × 150 km, and the resolution is 1—1.5 km. A similar picture can be taken from the Moon with the aid of a telescope with a field of view equal to 1.5′, and whose angular resolution is of the order of 0.9″.

The next photograph (Fig. 4) was taken from the "Nimbus" satellite by a T.V. system with a comparatively low resolution. The field of view embraces a plot of the Earth's surface 1,800 × 1,800 km in dimensions (spatial resolution is about 2 km). While carrying out the observations from the Moon the analogous image can be obtained by use of a telescope with a field of view equal to 16′ and an angular resolution equal to 1″.

The last photograph is that obtained with the system of the "Tiros I" satellite from an altitude of 627 km with an angle of view of 70° and a spatial resolution of 15 km (Fig. 5). The corresponding case of observing from the Moon would be obtained with a telescope whose angle of view is equal to 8′ and whose angular resolution is 4.2″.

Now, consider the concrete techniques through the use of which one can study the distribution of cloudiness over the globe.

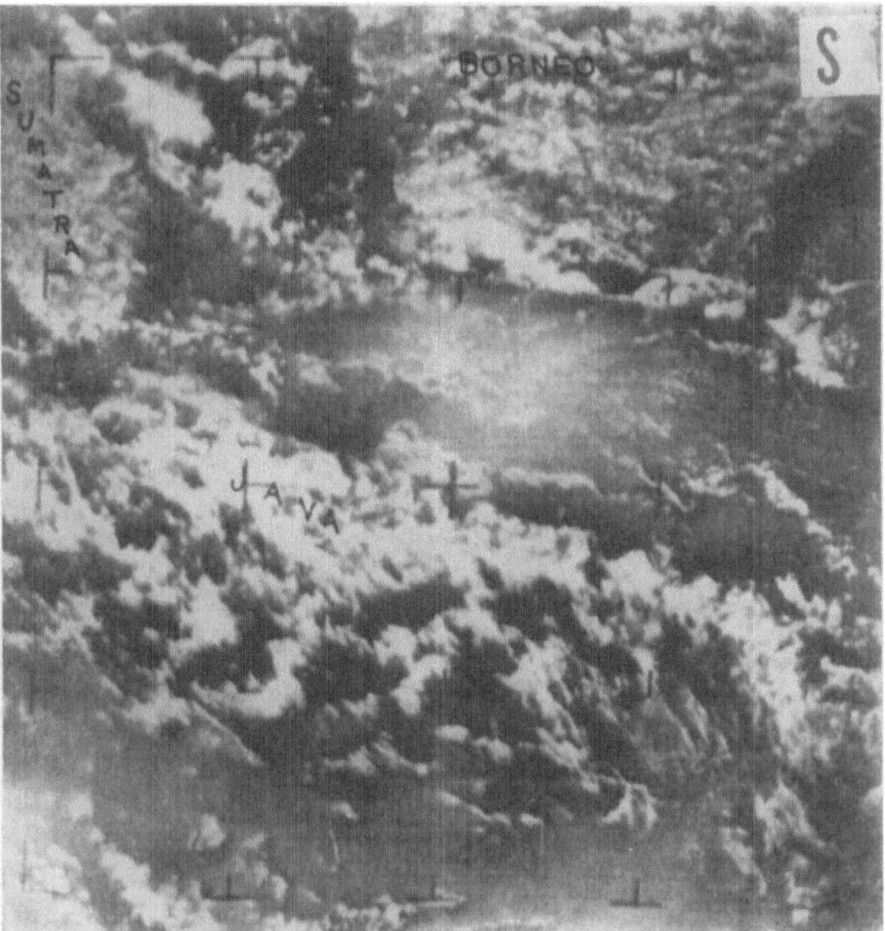

Fig. 4. Photograph obtained from the "Nimbus" satellite with a resolution of 2 km

First of all, there is T.V. tracking of cloudiness, the aim of which is to determine the distribution of cloud cover in daylight. Here, the principal task is to find optimal characteristics of the T.V. equipment which would permit the solution of a stated problem.

According to ERICKSON's criterion [1], the distance between the lines of a T.V. picture must not exceed 1.5 km on the Earth's surface. Thus with a picture containing 625 lines the T.V. camera's field of view must not cover an area more than 900 km in diameter. Though such resolution excludes the possibility of observing Cumulus humilis, their dimension being smaller than 2 km, still such a system will enable us to solve most of the meteorological problems dealing with the detection and tracking of cloudiness. Therefore, it will be necessary to design the T.V. camera with the angle of view in the order of 12′.

As for the brightness of the object — the Earth — no difficulties are expected, for it considerably surpasses the sensitivity of modern T.V. transmitting tubes.

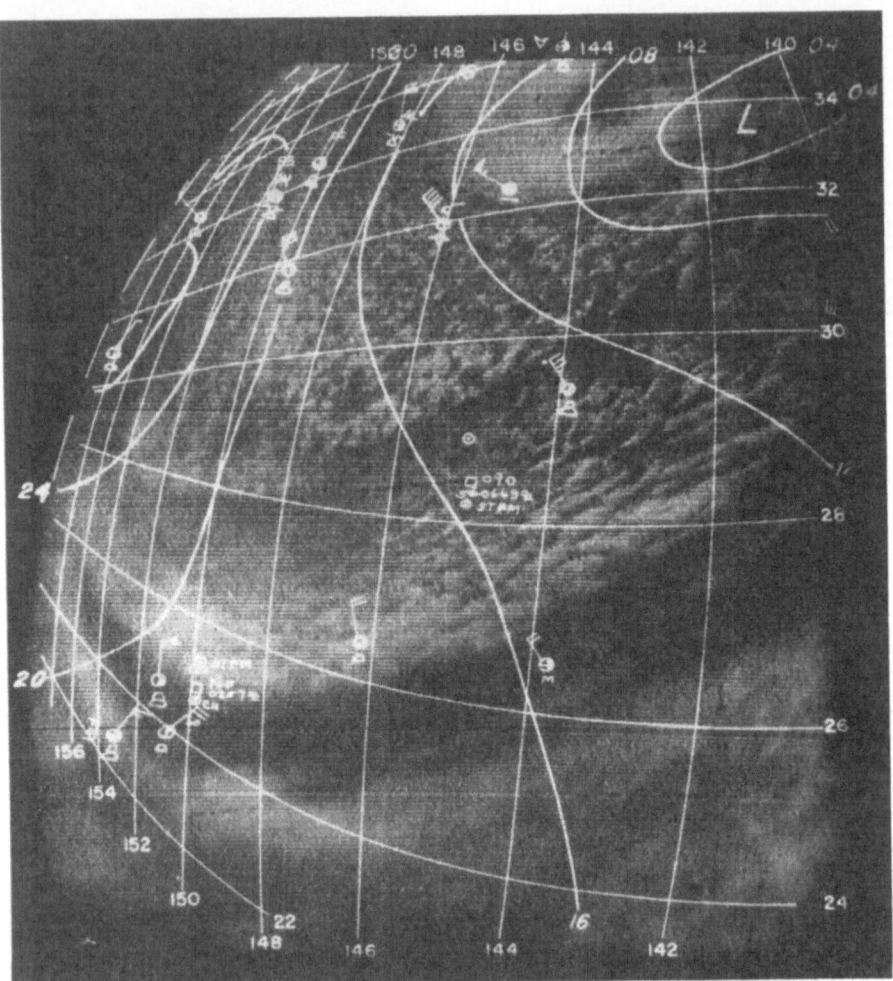

Fig. 5. Photograph obtained from the "Tiros I" satellite with a low spatial resolution

It is obvious that to obtain a complete picture of cloud distribution over the Earth's surface several T.V. cameras must be used, so that they overlap, even if partially, the surface of the Northern hemisphere. Further synchronisation of the pictures obtained will probably necessitate the application of one or several cameras with wider angles.

Investigation of Spatial Distribution of Cloudiness by the Method of Infrared Images

Investigations conducted in recent years show that the spectral regions of $10.5-11.2 \mu$ and of $3.5-4.2 \mu$ are the most suitable for this purpose. We calculated the intensities of the Earth's outgoing radiation at the level of the lunar surface for the above mentioned spectral ranges, as well as for temperature intervals of $10° K$ from $270° K$ and $260° K$. These intensities, taking into consideration the law of inverse root squares, have the following values:

$$10.5-11.2 \mu: E_1 = 3.1 \times 10^{-9} \text{ W/cm}^2,$$

$$3.5-4.2 \mu: E_2 = 18.5 \times 10^{-11} \text{ W/cm}^2.$$

In the examples considered the calculations have been made for light filters with 100% transmission. If one allows for losses in the optical system, in filters, etc., the above given values will be about half as much. Thus we have:

$$E_1 = 1.5 \times 10^{-9} \text{ W/cm}^2 \text{ and}$$

$$E_2 = 9 \times 10^{-11} \text{ W/cm}^2, \text{ respectively.}$$

Since the sensitivity of modern thermal radiation detectors does not exceed several hundred volts per watt, an optical system for the concentration of a bundle of light is needed to perform such measurements. Evidently, in the infrared spectral region, only a mirror system will be suitable. It is necessary to assure the discrimination of an area at least 200×200 km in dimension. Thus an angular resolution of $2'$ will be required.

Investigations of Cloudiness and Precipitation by Means of Passive Radio-Location

Methods of passive radio-location for the detection of precipitation and cloudiness zones have been rather well developed during recent years. New lightweight designs of pneumatic and folding radar systems possessing a very narrow main lobe of the direction diagram, up to several seconds, and with the diameter of opening equal to 10 m, can be successfully employed in a lunar observatory.

Let us consider the possibility of utilizing this equipment on the Moon. An analysis of radiometer equations allows us to conclude that the range of detection is most strongly influenced by the area and apparent temperature of the object under investigation as well as by radiometer quality. Scanning speed and effective signal duration are less important. The influence of the amplification factor is not strong, and it depends both on the conditions of visibility and on the relative dimensions of the object being observed.

Hight contrast (up to $100° K$) in the apparent temperature of underlying surfaces such as water and land or of Cumulonimbus ($50° K$) clouds makes radioemission measurements from a Moon observatory very promising.

Investigation of the Heat Balance of the Earth's Surface-Atmosphere System

At present we have quite reliable data on the radiation balance of the Earth's surface — atmosphere system. Its seasonal and latidunal curve is known; we have also information on the Earth's albedo, which has been obtained by a number of indirect methods. Moreover, with meteorological satellites of the "Tiros" and "Nimbus" series, direct measurements of outgoing radiation have been taken. But the problem of continuously recording the components of the Earth's heat balance, which are a main factor of the heat regime and the dynamics of the atmosphere, remains unsolved.

As is known, the equation of the Earth's heat balance has the following expression:

$$R_s = Q_0(1 - A_s) - F_\infty$$

where Q_0 is the incoming solar radiation. A_s is the albedo of the Earth as a planet and F_∞ is the outgoing long-wave radiation.

Since the solar constant, Q_0, can be recorded from the Moon at any moment of time with a sufficiently high accuracy (the existing apparatus ensures an accuracy of about $\pm 0.2\%$). Q_0 is not difficult to determine. The same applies also to measurements of albedo and outgoing radiation.

3. Recording of Solar Activity

One of the principal problems of a lunar meteorological observatory will be to find the relationship of solar activity to the processes in the upper and lower parts of the terrestrial atmosphere. It is apparent that in the thermosphere nearly all the processes are related to solar activity, since the energy of the processes is comparable in value to the changes in the energy flux determined by the changes in solar activity. Whereas, in the troposphere, the energy of any cyclone is by many orders of magnitude larger than the energy of the largest solar flare absorbed by the entire atmospheric thickness. Therefore, it might seem that the development of dynamic processes in the troposphere must not be associated with solar activity. However, a great number of works prove just the opposite. It was discovered that the cycles of solar activity were correlated with a variety of phenomena such as atmospheric meteorological characteristics, the number of earth-quakes, the levels of some water reservoirs, the growth of trees, etc.

There is a large number of contradictory results that arise from the application of different techniques for processing the data, as well as from choosing different indices for characterising solar activity. Nevertheless, at present, we cannot deny the existence of a relationship between certain tropospheric processes and solar activity.

One of the most important tasks facing meteorology is the study of the mechanism of this relationship. Some hypotheses explaining the mechanism of the effect of solar activity on weather-forming processes have been suggested, but because of drawbacks none of them can give an explanation of the relations observed.

For the solution of the problem being considered one needs to know, first of all, the following:

1) at which place of the terrestrial atmosphere the disturbance caused by solar activity occurs, — the disturbance is responsible for changes in the tropospheric circulation;

2) at which moment the disturbance arises;

3) the dependence of the character of atmospheric disturbances on the amplitude and duration of solar activity.

The chief condition for the solution of these questions is a continuous recording of solar activity in the whole spectrum of electromagnetic and corpuscular radiation, together with observations of the atmospheric reaction to solar disturbances. A continuous record will permit us to separate corpuscular effects from electromagnetic ones, and to estimate the effect of corpuscles with different energies at the moment of their interaction with different atmospheric layers.

Furthermore, a continuous record will make it possible to correlate terrestrial atmospheric changes directly with the energy and duration of a disturbance instead of with an arbitrary index of solar activity.

It is evident that the Moon's surface is the most suitable place for carrying out such correlation work. The large distance between the Moon and the Earth will enable us to make use of the difference between the time of interaction of the solar energy flux with the Earth and the Moon.

The lack of an atmosphere on the Moon and its low speed of revolution offer the best conditions for conducting these observations.

The lunar observatory may also be of great use in seeking new and deeper correlations of a planetary character. The attempts to find correlations between the Earth's thermal radio-temperature and the solar-activity cycle is of particular interest. The presence of such correlations would indicate a common cause of periodicity of the Earth and the Sun rather than an effect of solar activity on the Earth.

4. Complex Meteorological Investigations with the Help of Satellites and of a Lunar Observatory

A system of meteorological satellites cannot be replaced by a lunar observatory as, first, the latter does not permit simultaneous observations over the entire globe, second, on account of the large distance between the Earth and the Moon, not all satellite methods of atmospheric investigations can be applied on the Moon. This emphasizes the necessity of a complex application of measurements made by both artificial satellites and a lunar observatory.

The function of meteorological satellites is to provide continuous meteorological information on the terrestrial atmosphere as a whole, with a good angular resolution, for the purpose of weather forecasting. The Moon, on the other hand, is a suitable place for the solution of climatological problems which need a smaller angular resolution, and for which a continuous observation of the entire Earth's surface is not essential.

A lunar observatory will, evidently, permit us to carry out separate soundings of the terrestrial atmosphere with a greater accuracy than from satellites. The results of such soundings can also serve to some extent as bench marks in interpreting satellite data.

The use of meteorological satellites and a lunar observatory will be very convenient for investigating the atmospheric portion occurring at a fixed moment between a satellite and the Moon. Especially, the moments when a satellite emerges over the horizon and disappears beyond the horizon are of particular interest.

5. Possibilities of Meteorological Soundings from the Moon's Surface

Some of the problems of this kind are discussed in Ref. [1] as regards the application to satellite meteorology.

Determination of the Vertical Temperature Profile

The satellite method of determining the vertical temperature profile is based on a study of the spectral distribution of the outgoing thermal radiation of the Earth's surface-atmosphere system, since this thermal radiation in different spectral intervals is determined by different atmospheric layers.

Measurements with instruments of high resolving power are most promising from the point of view of obtaining a variety of data which can be relatively simply interpreted. In this case the problem of estimating the vertical temperature distribution by measurements of the CO_2 lines is expressed by a Fredholm integral equation of the 1st kind. Until recently the solution of such an equation was considered to be very difficult. Now great achievements have been made in the solution of non-correct mathematical problems (in particular, A. N. Tichonov, N. M. Lavrentyev and other Soviet mathematicians have elaborated some effective methods for solving linear and non-linear equations of the 1st kind).

Thermal soundings of the terrestrial atmosphere from the Moon would not have any special advantages as compared to satellite soundings. The large distance between the Earth and the Moon considerably decreases the energy flux received by an instrument on the Moon compared to the flux received by a meteorological satellite detector, with the spatial resolution from the Moon and from a satellite being the same.

However, in principle, there exists the possibility of conducting thermal soundings from the Moon.

The radiative flux near the 15 μ wavelength of a spectral interval 5 cm^{-1} in width, received from an area of the Earth $200-250$ km in diameter by a unit surface of a lunar detector has the value of the order of 10^{-11} w/cm^2. If the surface of the receiving device is sufficiently large, this flux can be recorded.

Determination of the Upper Boundary of Cloud Cover

Passive methods of determination of the upper cloud boundary, based on optical methods for measuring the amount of the absorbing gases (CO_2, H_2O or O_2) above clouds, are not suitable for use by a lunar meteorological observatory for the above mentioned reason. That is because of the loss in energy due to the greater Moon-Earth distance, with the same angular resolution compared to an artificial Earth Satellite. Active light-location and radio-location methods seem to be more promising under the conditions of a Moon's observatory, assuming that it can be supplied with energy to the extent of an observatory on the Earth.

Determination of the Ozone Concentration

Meteorologists take an interest in ozone by virtue of the important part it plays in the heat regime of the stratosphere, made evident by the presence of strong absorption bands in the ultra-violet and infrared spectral regions. Considerable time and spatial variability in ozone concentration, which has no theoretical explanation up to now, affects stratospheric circulation, and through it the weather-forming processes in the troposphere. Since measurements of ozone are taken only at a limited number of points, they cannot be considered as fully reliable meteorological data. From this it follows that continuous global observations of ozone are needed for improving (or perhaps, correcting) the photochemical theory of ozone formation, as well as for purely meteorological purposes, in the narrow sense of the word.

The elaborate optical methods for obtaining satellite information on the ozone layer have the same drawbacks as land optical methods. The difficulty of interpreting these results is connected with the necessity of taking into account the variability both of the ultra-violet solar emission and of the transparency of the terrestrial atmosphere on the one hand, and with the influence of aerosol and multiple scattering on the other. The application of these methods in a lunar observatory is much more difficult than from satellites, except that to overcome the above mentioned difficulties it is necessary that the measurements of very small radiative fluxes be ensured.

The application of active methods for the investigation of the ozone layer from the Moon would be more feasible, if these could ensure greater accuracy as compared to passive ones.

Study of Atmospheric Horizontal Optical Non-Uniformity by the Method of Laser Sounding

The laser is likely to become the principal means for active soundings of the terrestrial atmosphere from the Moon. The main advantages of employing the laser as the means for making soundings are the following:

1) narrowness of the laser bundle makes it possible to get a large spatial resolution;

2) the great impulse power and the high degree of monochromaticity enable us to single out laser emission against the background of cosmic objects and, in particular, against the background of direct solar light;

3) the use of simple equations for monochromatic emission transfer ensures comparatively easy interpretation of experimental results;

4) there are some possibilities of constructing lasers for operation in different spectral regions.

Consider an impulse laser emitting at a wavelength of 6,000 Å in a band 10^{-5} cm^{-1} in width, with a power of 10^9 watts and with an impulse duration of 5×10^{-8} sec. With a lens 5 m in diameter one may obtain a maximum value of the direction of the laser bundle equal to 10^{-7} rad. While emitting from the Moon such a laser will produce on the Earth a spot whose diameter is about 40 m and whose brightness (with diffuse reflection and albedo equal to 0.3 and with the atmospheric extinction as much as three times) of about 3 W/cm^2 sterad, which is much greater than that of the Sun for the narrow spectral region within which the laser is used. If one employs the laser impulse for a double passing through the distance between the Earth and the Moon, the 5 metre receiving mirror on the Moon will receive the impulse weakened as much as 10^{-18} times. So, about 100 photons will reach the mirror. The level of background from the sunlit part of the Earth in the spectral interval considered will be much smaller than 100 photons, if the angle of view of the detector cutting out a small part of the Earth's surface is sufficiently small.

Thus, the reflected impulse may be recorded. If lasers of continuous action with the power of 1—10 kW are designed, their use will be still more promising.

The use of sounding lasers from a Moon observatory is more convenient than on satellites because on the Moon we shall not face the problem of sizes, feed sources, etc.

Our considerations testify to the great role a lunar meteorological observatory can play in improving weather and climatic forecasting.

References

1. K. Y. Kondratiev, Meteorological Satellites, p. 261. Moscow: Gidrometeoizdat, 1963.
2. V. V. Shevchenko, Visible Displacements of the Earth in the Lunar Sky. Cosmic Investigations 1, Issue 2, p. 216 (1963).
3. K. Ya. Kondratiev, M. G. Kroshkin and V. G. Morachevsky, Our Planet, from Space, edited by E. K. Fyodorov, p. 30. Moscow: Gidrometeoizdat, 1964.

Lunar Astronomical Observatory

By

Kurt R. Stehling[1] and I. M. Levitt[2]

(With 17 Figures)

Abstract — Résumé — Резюме

Lunar Astronomical Observatory. The authors review briefly the advantages of, and need for, an astronomical observatory on the moon. They review the optical advantages of an instrument operated in the airless environment of the lunar surface. Also outlined are reasons for proposing a telescope of a sufficient aperture to yield new and significant astrophysical and astronomical information. They consider the lunar observatory to be centered on a reflecting telescope of no less than 150 cm aperture and perhaps as much as 250 cm aperture.

The authors then treat in some detail the logistic and mechanical problems of transporting a telescope and its accessories to the moon in the first place, by the LEM or a similar spacecraft. New information is given on light-weight optical development, some of which is in an advanced state with capability approaching the manufacture of astronomical quality reflecting surfaces.

The structural, assembly and alignment problems of such an instrument are discussed as well as the virtues or defects of new methods of light-weight mirror fabrication. Described are techniques for electro-forming metallic mirrors, the buildup of quartz mirrors and the possibility of fabrication of optical elements on the lunar surface, including such techniques as the spinning of liquid metal pools to achieve astronomical reflecting surfaces.

Various types of telescope and observatory constructions are considered including the use of coelostats and other fixed optical devices. Also, consideration is given to the establishment of photoelectric and video telescopes on the lunar surface, as well as the problems that are likely to arise from the standpoint of manipulating and general operations.

The kind of important astronomical information that may be obtained from a lunar observatory is summarized. A number of illustrations of possible light weight optical elements and telescope configurations are included.

Observatoire Astronomique Lunaire. Les auteurs rappellent rapidement les avantages et la nécessité d'un laboratoire astronomique sur la lune. Ils passen en revue les avantages optiques d'un instrument utilisé dans l'environnement dépourvu d'air, à la surface de la lune. Ils esquissent les raisons d'un projet de télescope d'ouverture suffisante pour fournir des renseignements nouveaux et importants en astrophysique et en astronomie. Ils considèrent que l'observatoire lunaire doit être centré sur un télescope catoptrique n'ayant pas moins de 150 peut-être 250 cm d'ouverture.

[1] Assistant to the President for Astronautics, Electro-Optical Systems, Inc., Pasadena, California, U.S.A.

[2] Director, Fels Planetarium of the Franklin Institute, Philadelphia, Pennsylvania, U.S.A.

Les auteurs traitent ensuite en détail les problèmes de logistique et de mécanique posés par le transport même du télescope et de ses accessoires sur la lune, à l'aide du LEM ou d'un engin spatial analogue. Ils donnent des renseignements nouveaux sur le développement des optiques à poids réduits, certaines d'entre elles étant dans un état assez avancé pour permettre de projeter la fabrication de surfaces réfléchissantes de qualité astronomiques.

On discute les problèmes de structure, d'assemblage et d'alignement de tels instruments de même que des qualités et des défauts des nouvelles méthodes de fabrication de miroirs légers. On décrit les techniques de miroirs métalliques d'électro-formation, la construction de miroirs de quartz et la possibilité de fabriquer des éléments optiques sur la lune, comprenant des techniques telles que la rotation de masses métalliques liquides pour réaliser des surfaces astronomiques réfléchissantes.

On considère divers types de télescopes et de construction d'observatoires, y compris les cœlestats et autres dispositifs optiques fixes. On envisage aussi sur la lune de télescopes photoélectriques et de télescopes à télévision de même que les problèmes qui résulteront probablement des manipulations et des opérations générales.

On résume rapidement le genre de renseignements importants qu'on pourrait obtenir à partir d'un laboratoire lunaire. On ajoute quelques illustrations représentant des éléments optiques légers et des configurations possibles de télescope.

Лунная астрономическая обсерватория. Авторы кратко рассматривают преимущества и необходимость создания астрономической обсерватории на Луне. Они останавливаются на оптических преимуществах использования прибора для наблюдений в условиях отсутствия воздушной атмосферы на поверхности Луны. Подчеркивают необходимость в телескопе достаточно большого диаметра, для того чтобы получить новые и важные астрофизические и астрономические сведения. Они считают, что главным элементом лунной лаборатории должен быть телескопрефлектор диаметром не менее 150 см, возможно даже до 250 см.

Затем авторы рассматривают организационные и технические проблемы, связанные с переброской на Луну телескопа и его оборудования с помощью космического корабля ЛЕМ или другого подобного ему. Приводится новая информация о достижениях в области создания легкой оптики — некоторые разработки находятся уже в стадии близкой к обеспечению производства отражающих поверхностей астрономического качества.

Рассматриваются проблемы конструкции, монтажа и установки такого прибора, а также достоинства и недостатки новых методов производства легких зеркал. Описываются методы электроформовки металлических зеркал, конструкция кварцевых зеркал и возможность изготовления оптических элементов на Луне, включая такие методы, как вращающийся сосуд с жидким металлом, обеспечивающий получение астрономических отражающих поверхностей.

Рассматриваются различные конструкции телескопов и обсерватории, включая использование целостатов и других неподвижных оптических приборов. Рассматривается также вопрос об установке фотоэлектрических и видео-телескопов на Луне и проблемы, которые могут возникнуть в связи с работой на приборах и общей эксплуатацией обсерватории.

Перечисляются виды важной астрономической информации, которая может быть получена благодаря лунной обсерватории. Сообщение содержит ряд иллюстраций, показывающих возможные варианты легких оптических элементов и конструкций телескопа.

Although we have barely accomplished an unmanned landing on the moon, and are some years away from a manned lunar landing, it is not too early to plan for the possible follow-on activities that may accrue from a first lunar manned landing.

Considerable thought has been given to the sorts of experiments and activities that could be performed on the lunar surface, both with and without men. Many space technologists who have been so sanguine as to believe that unmanned lunar research apparatus could be set up on that surface have realized that a man would be a necessary and useful addition to any lunar operation despite the great problems of transporting him and keeping him there. This will not deny the fact that a modicum of unmanned exploration will be possible and necessary and that data about the lunar terrain and environment will be learned through instrument packages. As long as the functions of such packages are fairly simple, unmanned operations can serve a useful and perhaps necessary role before man lands.

If it is found after the first landing or two that men with the proper shelter and protection can survive the undoubtedly inimical environment on the lunar surface, what experiments and activities could be carried on to the benefit of science and space technology? A number of activities have been proposed for manned operations on a lunar base. These activities include geology, mineralogy, seismology, space radiation experimentation, biophysics and bioastronautics, and astronomy. It is the latter area that we wish to deal with in more detail in this article.

The Moon provides an ideal platform for the location of astrophysical instrumentation primarily by virtue of its, hopefully, complete freedom of an atmosphere. This permits unhindered optical viewing of the various heavenly objects, free of atmospheric absorption, distortion and scintillation—the bane of earth-bound astronomers. Spectroscopic research, the mainstay of astrophysics, can be now utilized in all its majesty from the microwave end of the spectrum to the gamma

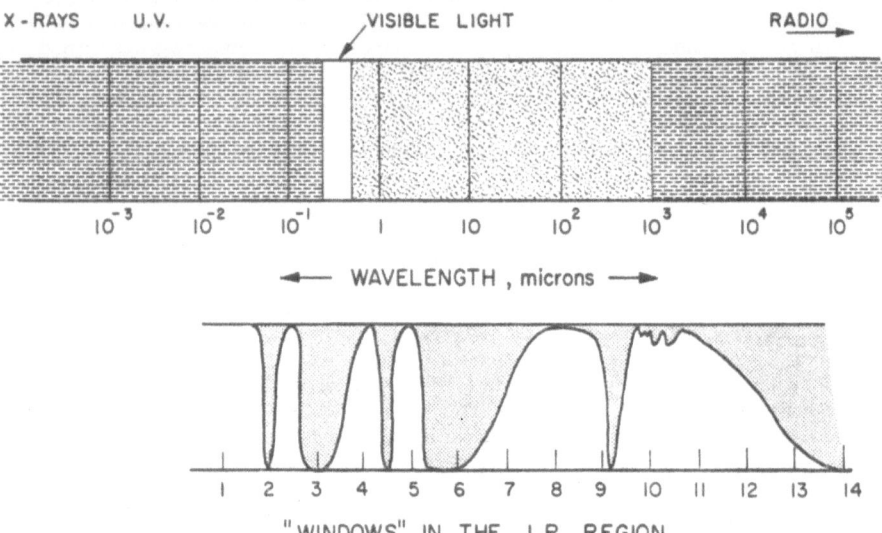

Fig. 1. Spectral transmission bands or "windows" in the atmosphere. Clear areas represent radiation transmission through atmosphere

ray region (Fig. 1). Also, the radio astronomer—now a very important member of the world's astronomical community and one that must surely be represented in such an endeavor—will be able to operate with great freedom from man-made or natural terrestrial electrical noise. There will be little of the present interference from assorted objects in Earth orbit including iron filing, empty cans and pots and missile junk, ionization noises from meteor tracks and radio emissions from spacecraft.

The next virtue that the Moon possesses for the astronomer is its relatively low gravity which permits the mounting of sizeable structures with less strain, bending and warping than occurs on Earth. For instance, the astronomical mirror is subjected to different stresses as the telescope moves. These are compensated for by various stratagems, none of them completely satisfactory.

The Moon offers an arena of undisturbed environment, quiet from the standpoint of electrical and mechanical noise and atmospheric clutter and absorption. We are not ignoring the hair-raising problems of moving material to the lunar surface and setting up equipment for the purposes of astronomy. (See Tables I and II.)

Table I. *Advantages of a Lunar Observatory*

1. No atmospheric absorption, or refraction across entire electromagnetic spectrum and of course no atmospheric interference of any kind. Visibility is same from horizon to horizon.
2. Long continuous observing time — almost 27 days at the poles, or at least 15 days if observing ceases during lunar day.
3. Slow drift of star field — about 0.5°/hr compared to 15°/hr on earth. Reduces guidance problems except for predictable laboratory motions.
4. Little or no stray or scattered natural or man-made light or other radiation is present.
5. Man-made seismic distribances are absent; hopefully, natural seismic tremors are much less frequent than such phenomena on the earth, where large thermal stresses in the earth's crust are common.
6. A stable and firm platform is available, offering shelter from meteors and radiation, including the sun. Operating and maintenance facilities are possible.
7. Some (1/6 of earth) gravitational force is present, eliminating or reducing problems of manned operation in "0" g.
8. Manned operation permits a wide variety of observations (e.g., photography, spectro-photography, interferometry, coronography, photometry) which are difficult, if not impossible, to achieve with a single unmanned earth orbiting telescope.
9. In-situ measurements and calibrations are possible, as well as analysis and interpretation of photographic and other data. Photographic film can be protected from "fogging" by stray radiation or bursts of solar or stellar radiation.
10. For solar observation a coronograph could be used to its full advantage because of the absence of atmospheric clutter or dust.

Table II. *Disadvantages of a Lunar Observatory*

1. The cost and logistic problems of transporting material and men to and from the moon.
2. The problems of assembling, operating and maintaining equipment in the hostile environment of the lunar surface.
3. Possible disturbances by dust particle scintillation near the lunar surface, especially during solar illumination.
4. Very high electronic and mechanical reliabilities will be needed to reduce maintenance and repair requirements.

Therefore, we will discuss briefly the probable limitations that face the astronomer who wished to package optical telescopes, radio telescopes and associated gear into the crowded confines of a lunar spacecraft as it may be launched by the Saturn V. The equipment that is packaged must be able to survive the rigors of the flight and the even greater rigors of the landing shock. All apparatus must be capable of being disassembled by a few men working in a inimical environment. We mentioned the term, "packageable". By the very nature of astronomical research and equipment used, rigidity and dimensional stability are not only important but necessary attributes of such equipment. As an example, a large astronomical mirror, say about 60 to 100 (150—250 cm) inches diameter or more, or for that matter any astronomical mirror, must not show perceptible flexure as it is swung about in a telescope tube. This disturbing flexure, if it should occur, can be compensated by exerting variable pressure against the back of this mirror and more importantly by rendering the mirror stiff by making thick and heavy or, alternatively, as light as possible with reinforcing ribs or cells. Similarly, the telescope body and mounting and all moving associated parts must be very rigid and therefore usually heavy and bulky.

The slightest asymmetry in the optical alignment of the telescope ruins its function. Also, external vibrations caused by traffic, etc., must be damped out. This again is generally done on Earth through the use of very heavy concrete and steel "massives" and emplacements and bases with pillars sometimes extending many meters into the ground or rock beneath. For example, the mirror, alone, of the 100 inch (250 cm) Hooker (Mt. Wilson) telescope weighs 9,000 lbs (4,050 kg), while the Mt. Polomar 200-inch diameter (500 cm) mirror weighs 29,000 lbs (13,000 kg). These are just mirror weights, and do not include the hundreds of tons of concrete and steel needed for the support pillars and mountings. (The Mt. Polomar telescope tube, mirror and observer's cage weighs 140 metric tons.)

Also, the telescope must have a precise mechanical motion to follow the celestial body it is tracking. This motion must be smooth, uniform and reproducible and must be ascertainable at all times. In an Earthbound telescope this is relatively simply done through the mounting of the telescope on a special axis one end of which points to the north celestial pole. Thus the telescope mounted on such an axis needs only one motion at a time to follow an object in the sky as this object moves across the heavens — this single motion is possible because the telescope is mounted on an axis which is parallel to the Earth's axis of rotation.

On the Moon, on the other hand, the situation is different. The Moon rotates synodically once in every 27-1/3 days with the same face always turned toward the center of the Earth. We presume that the observatory is located on the "far" side of the Moon to eliminate the effects of "Earth shine," although the latter would be much less a problem than "Moon shine" on earth, because of the absence of an atmosphere. If a telescope is mounted at other points except the lunar "poles," it will be possible to see a given portion of the sky for periods of time not exceeding two weeks. The small diameter of the Moon yields a fairly "close" horizon for the astronomer although this is not a particulary serious objection.

The star field drifts across the lunar sky about 0.5°/hr as compared with 15°/hr on the Earth. This is "rotation" drift. The proper motion drift or position as a function of the date is given on Earth by time-referenced right ascension and declination. A different, but easily attainable system could be arranged using Earth "real-time" reference or a Moon-based and referenced clock.

However, the type of mounting and the method of tracking a celestial body will have to be different from that in vogue on Earth, and the location of the telescope will have to be in an area that will permit the maximum amount of seeing. Of course, the location of an observatory cannot be entirely determined on the seeing factors alone because of the problems of establishing a lunar base in an area accessible to rocket landings and relatively free of geological anomalies and features such as deep craters, fissures or crevices. A similar situation holds on Earth — seeing might be considerably improved from the standpoint of atmospheric absorption and scintillation if a telescope could be moved to a 20 or 25 thousand foot high (6,000—8,000 meter) mountain provided there were, of course, no problems of snow or high velocity winds. That observatories have not been built on such high mountains is attributed to the common sense of the engineers who very likely would rebel against a task so gargantuan, to say nothing of the constant wheezing and breathing difficulties that the astronomers would have in an atmosphere above 20,000 feet.

The above statements do not only apply to an optical observatory and its associated equipment, but also to a radio observatory. A radio telescope will not require the same sophisticated housing that optical telescopes need, but on the other hand it does require more terrain (lurain) because of the much bigger diameter of the radio antenna, and the requirements of feeding the signal from the antenna to a receiving station which can be at a considerable distance from the radio antenna. The dome so common for Earth-bound optical telescopes is probably not so necessary for the lunar telescope for various reasons which will be described later.

Let us consider the question of packaging the optical or radio telescope elements needed for the observatory. It must be remembered that the telescope and its accessories are still now and will probably continue to be, the primary instrument for astrophysical reasearch. For both the optical and radio case, the primary light or radiation collector — the mirror or the antenna — is the most important element and the most difficult and critical one to transport with sufficient physical integrity to permit erection on the Moon with reliable long-term operation. A question that must be raised here is — what aperture is practical, useful or necessary for a lunar operation?

In the optical case it is a fact *that any* telescope of any diameter will be greatly improved in its resolving power if it is transported to the Moon — as compared to what it can do on Earth. The same is not quite as true for associated equipment such as a spectrograph, where reception, say in the far IR or UV regions, might still be limited by the spectroscope's optics or the sensitivity and response of a photographic plate or detector. Nonetheless, a sizeable gain in information gathering ability would still accrue with even the simplest spectrographic or similar equipment.

A mirror or lens is necessary for the primary light collector. The mirror has obvious advantages over a lens not the least of which is its potentially lower weight and therefore cheaper transportation to the Moon.

To illuminate this point let us first review the volume and weight carrying capabilities of the LEM. (The designation of the U.S. "Lunar Excursion (i.e., landing) Module.") We presume here that the LEM or a variation of it, will be the only lunar supply transportation vehicle available for several years after the first manned lunar landing. Thus any early lunar observatory would depend on the LEM as a "truck". Now, the approximate instrument payload weight that can be carried to the Moon is about 7,000 lbs (3,000 kg) if a non-returnable payload-carrying unmanned module is assumed. This figure could be increased by about 10% if high energy propellants such as fluorine and hydrogen were used for the retro stages. If the LEM itself were used (i.e., the normal manned module, returnable) then only about 120—150 lbs (50—70 kg) could be available for astronomical purposes and mirror diameter of one meter or so, could be accommodated. If we assume that the LEM configuration will dominate the design of the "truck", then we can carry a flat object about 5 ft (1.5 meters) in diameter. This gives the limiting geometry for the mirror. An unmounted quartz or pyrex mirror of this size would weigh about 3,000 lbs (1,400 kg). It is apparent, therefore, that such a conventional mirror is difficult to transport to the Moon, or at least, that it could only be transported there to the exclusion of much other payload.

Thus a "conventional" mirror greatly increases the cost of a lunar telescope. It is necessary to be concenred with economics, especially since a scientific instrument or laboratory on the Moon cannot be used as readily for rationalization of high launch costs as much as a first or subsequent manned landing can be rationalized.

We face the question then of whether we can reduce the cost of lunar transportation. We can decrease the weight of the mirror sufficiently to bring its cost

Fig. 2. Primary mirror mounting used on the Orbiting Astronomical Observatory (OAO)

per lb down to a low and "reasonable" value. A parallel problem is breaking loose from the vehicle limits of 1 meter for the mirror's diameter. A smaller mirror would be considerably less than optimum or what is desirable for an established observatory, although a smaller aperature would perforce, have to do for initial and exploratory experiments.

We can reduce the weight by a thorough application of certain areas of advanced lightweight structure technology. An example is the beryllium mirror used on the OAO (Orbiting Astronomical Observatory). (Fig. 2, 3, 4 and 5.) This 40″ diameter mirror weighs about 120 lb (50 kg). For the lunar situation we should want the mirror to be even lighter per unit area. If we extrapolate the OAO mirror to the diameter of 100″, the weight, assuming it is the same thickness, would be about 750 lbs. This is still heavy. We could reduce the weight by changing the thickness, but the mirror might suffer a damaging flexure during rocket acceleration or Earth "g". If these factors could be obviated, however, a thickness reduction would be feasible, considering the 1/6 lunar "g" force. A material change does not look promising since beryllium is the lightest material we can hope to use, practically.

One way around this dilemma might be to use techniques presently being developed for forming, by applied pressure or spinning, the necessary parabolic

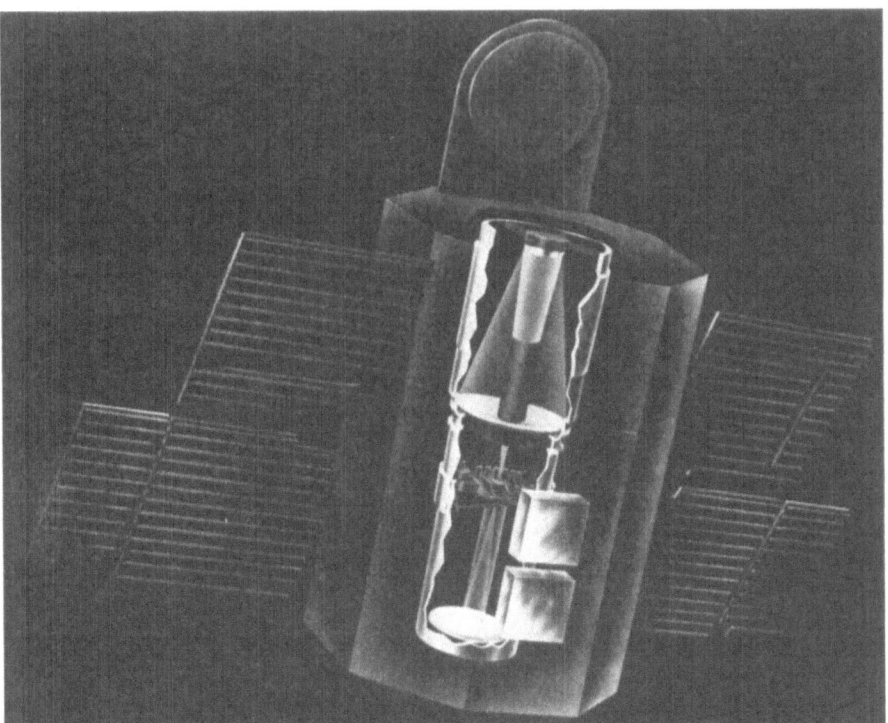

Fig. 3. Goddard experiment package for the Orbiting Astronomical Observatory (OAO)

Fig. 4. 36″ diameter $f/5$ Goddard experiment package optical system for the Orbiting Astronomical Observatory (OAO)

shape of our mirror using very thin sheets of material such as nickel, cooper, or perhaps aluminum or beryllium, and backing this mirror with certain plastic or quasi plastic substances of considerable low unit weight. (Fig. 6.) However, this technology is still in a relative infancy, and years of development are necessary to produce mirrors and other optical elements of astronomical accuracy.

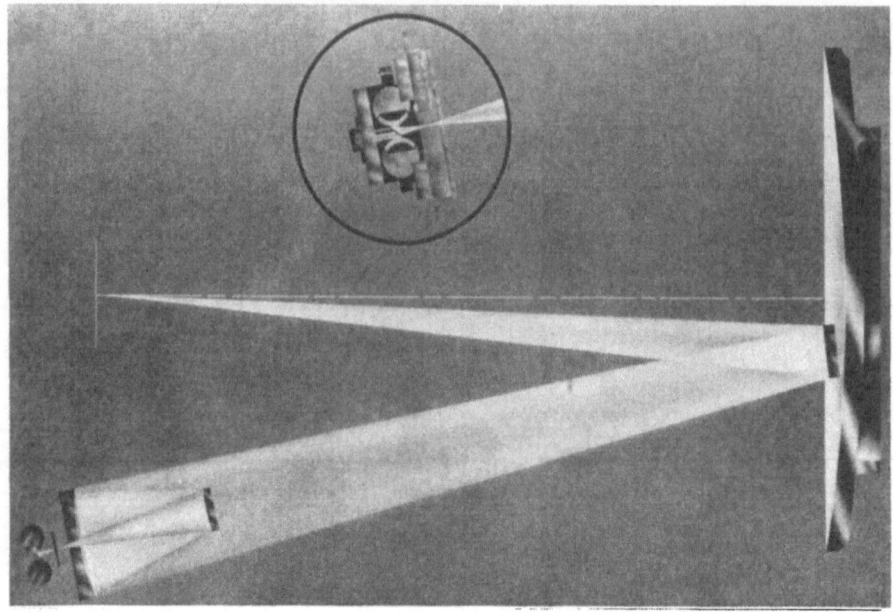

Fig. 5. Fine guidance subsystem of the Goddard experiment package for the Orbiting Astronomical Observatory (OAO)

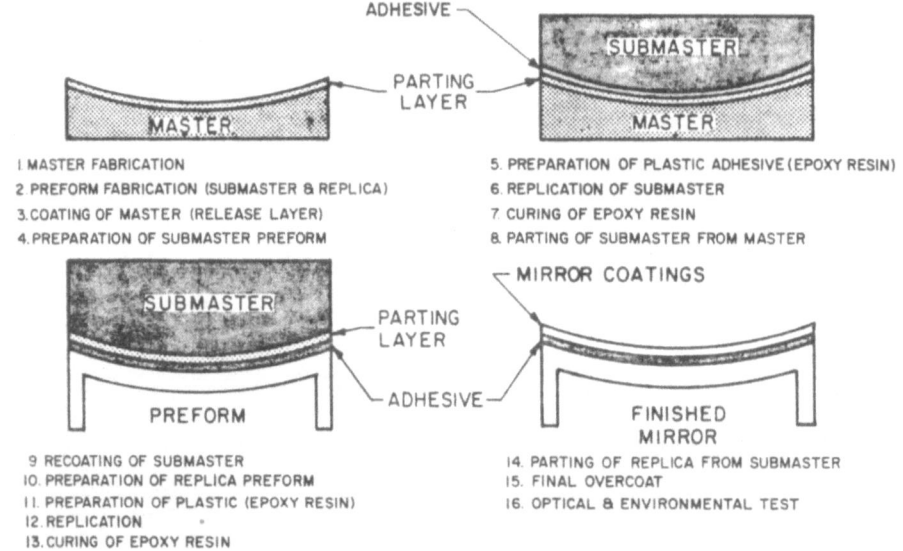

1. MASTER FABRICATION
2. PREFORM FABRICATION (SUBMASTER & REPLICA)
3. COATING OF MASTER (RELEASE LAYER)
4. PREPARATION OF SUBMASTER PREFORM

5. PREPARATION OF PLASTIC ADHESIVE (EPOXY RESIN)
6. REPLICATION OF SUBMASTER
7. CURING OF EPOXY RESIN
8. PARTING OF SUBMASTER FROM MASTER

9. RECOATING OF SUBMASTER
10. PREPARATION OF REPLICA PREFORM
11. PREPARATION OF PLASTIC (EPOXY RESIN)
12. REPLICATION
13. CURING OF EPOXY RESIN

14. PARTING OF REPLICA FROM SUBMASTER
15. FINAL OVERCOAT
16. OPTICAL & ENVIRONMENTAL TEST

Fig. 6. Steps in the adhesive-plastic replication process

Another very interesting technique, "electroforming", offers a promise of even lighter-weight mirrors; however, a long and expensive development program would also be necessary to produce astronomical optics. The best that has been achieved so far is the production of a 10 ft mirror weighing about 30 lbs (20 kg) with an accuracy of about 3—5 minutes of arc. This is a long way from the accuracy that is needed for an astronomical mirror. However, various forming and replicating techniques are at least a beginning in the domain of the manufacture of lightweight reflective surfaces. While it is not yet apparent how astronomical quality mirrors can be made by such processes, there is no doubt that refinements or variations in these techniques will produce such mirrors in another decade or so when we should be ready to start building an observatory on the Moon.

Let us return to the question of diameter or overall size. Assuming, by some forming and backing technique, we have brought the weight down to an accept-

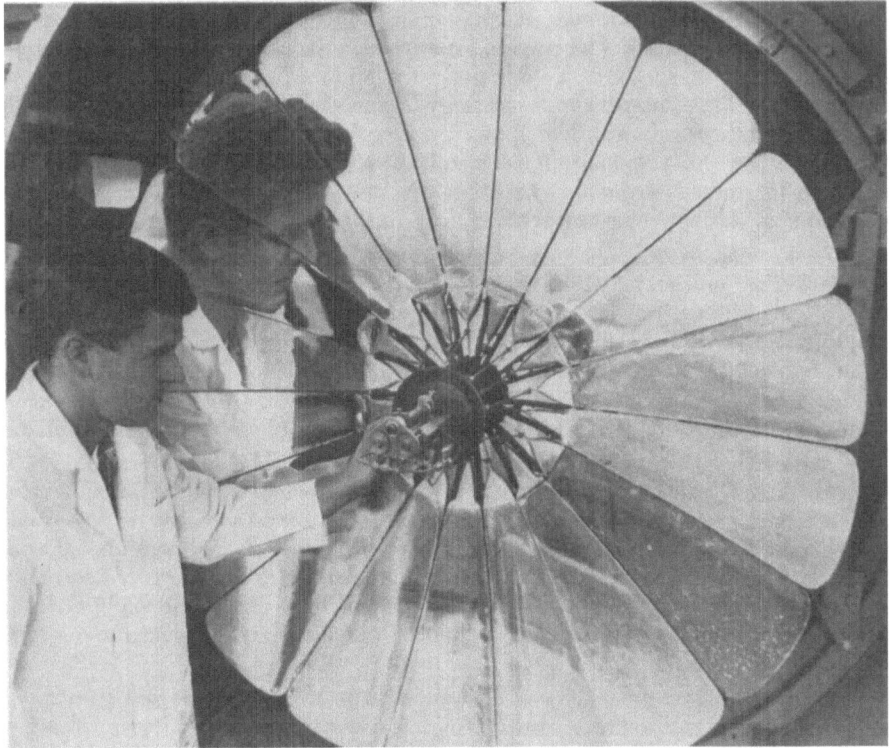

Fig. 7. Flower petal solar reflector (mirror)

able level, how can we accommodate a larger diameter mirror (say 3 meters) in the restricted quarters of a LEM "truck" without building a specific lunar lander for the mirror alone ? Again, we can return to this advanced mirror forming technology to give us a clue. We begin by making the radical assumption that ahe mirror may not need to be shipped to the Moon as a single entity. We can tsk why not send it up in pieces, and then assemble it, or why not build it like the collapsible reflector of miniature photo flash systems. Here we could have a flower petal arrangement that would permit the petals to open when desired, fall adjacent to each other, and then form one contiguous series of elements adding up to a mirror. Such a notion is hair-raising to the "classical" optical scientist and astronomer. These reactions are well founded if we base this notion on present technology.

Large flower petal solar reflectors (mirrors) have been built and used (see Fig. 7), and have done quite well. The NASA and the U.S. Department of Defense have been, and are interested in certain types of solar collectors and other reflective surfaces that would have to be transported in the restricted domain such as the Atlas-Agena, the Titan or even the Saturn I. However, the optical demands on such folding or collapsible mirrors would be considerably *less* from the standpoint of achieving a discrete and singular focal point or region such as would be needed for astronomical purposes, i.e., imaging. This whole notion then boils down to the question of achieving the necessary accuracy of unfolding so that all the leaves fall exactly into place without distortion and with an almost total absence of visible joints. Again, this is a very long way from the accuracy mentioned above that is needed. However, it is possible to be optimistic, and state that the technology of segmented and collapsible mirrors is in its infancy, and with time, say in a decade, we should be able to find ways and means of achieving astronomical mirror quality. The present technology is considerably ahead of the first crude attempts of a few years ago which had three or four times the errors of the latest mirrors. At any rate, such collapsible reflectors are even at present sufficiently accurate for radio astronomy (i.e., antenna) use, and thus could be the prime collectors of a radio observatory.

It is interest to note that there are some other possible methods of meeting the packaging and transportation problems of large mirrors. The idea is simply to fabricate the mirror on the Moon. Experiments by R. W. Wood at Johns Hopkins University in 1908 and in Germany in the late 1920's showed that a pool of mercury, when rotated, would attain a parabolic form of astronomical quality whose focal length varied with the speed of rotation. It was found difficult to eliminate vibrations that disturbed the surface and figure of the liquid mirror.

On the other hand, under good conditions of seeing and low vibration, a 20″ (50 cm) mercury mirror with a 20″ (50 cm) flat yielded excellent lunar images and resolved stars 3 second apart. Mercury could be transported to the Moon to become the required mirror. Assuming a mirror "pool" 2″ (5 cm) high, 5,000 lbs (2,200 kg) of mercury would be needed for an 80″ (200 cm) mirror. Of course, a turntable would have to be assembled. This could be assembled or built up in a fairly commonplace engineering fashion from elements transportable rather easily to the Moon.

The mercury is not as light as we would wish, and we would not gain much weight saving, but a definite gain in volume packaging accrues. Room for a few bottles or a tank of mercury could be fairly easily arranged in a LEM truck. Mercury is presently shipped in steel flasks. Such flasks must be made strong to prevent deformation by the heavy mercury. However, for our lunar use, it could be shipped probably in fiberglass or other lightweight containers which could be used for other purposes later on.

It is not necessary to consider only mercury — other substances could be used, such a liquid alkali materials with a low melting point or even the lighter materials such as beryllium or aluminium. There would be no serious thermal problem on the Moon if the astronomer were not interested in infrared measurements.

An optical flat or flats for a coelostat or siderostat would be necessary. However, the transportation, weight, optical and mechanical problems of such tracking systems would be relatively small compared with those of the primary mirror. The latter should be mounted in a deep well or pit to eliminate thermal problems. There would presumably be no man-made or natural seismic vibration.

Another method for forming a parabolic surface would be to subject an optically flat piece of material of glass or metal and relatively thin (a half inch to an inch, 1—2 cm, or so) to some type of symmetrical loading which would deform

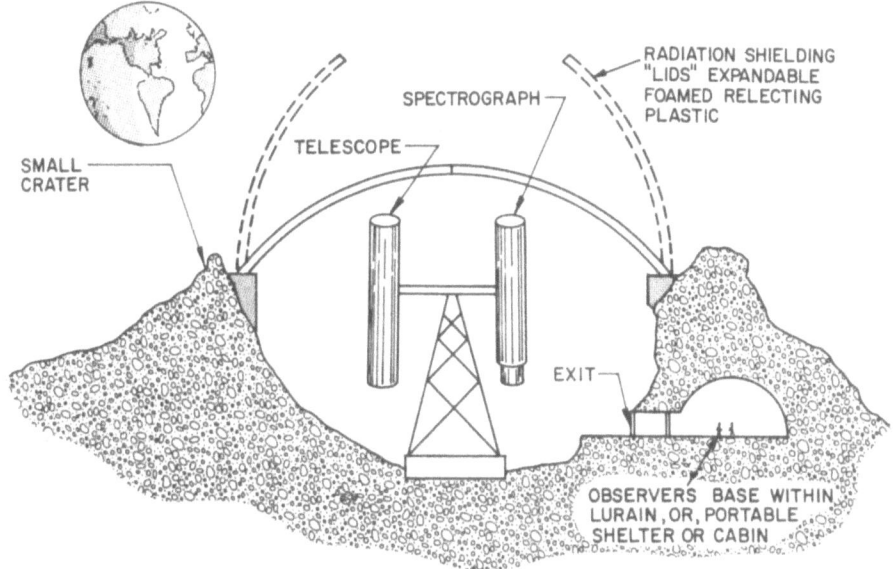

Fig. 8. Proposed "semi-permanent" observatory

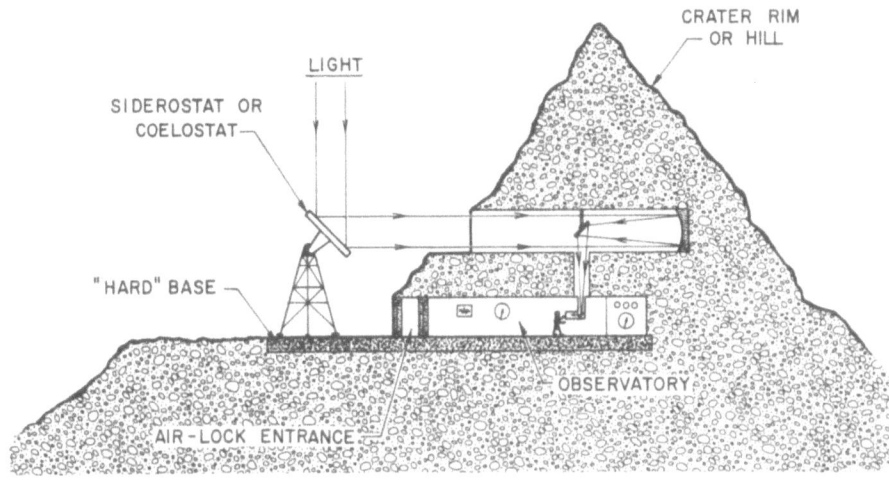

Fig. 9. Other "fixed-base" proposed observatory, protected from meteors and radiation

this sheet, cut in the form of a circle of the requisite diameter. One method of achieving such loading, and one that has been tried in laboratories, is to place the flat sheet over a cavity, and evacuate the cavity to a partial vacuum. Then the air pressure across the sheet would distend it into a shape approximating a parabola. On the Moon, of course, the reverse would have to be done, and pressurization would have to be applied to one side of the sheet. Other notions that have been suggested include loading this disc with a homogeneous material such as water which again would tend to bend the disc toward the center. The few laboratory experiments that have been tried show that with sufficient refine-

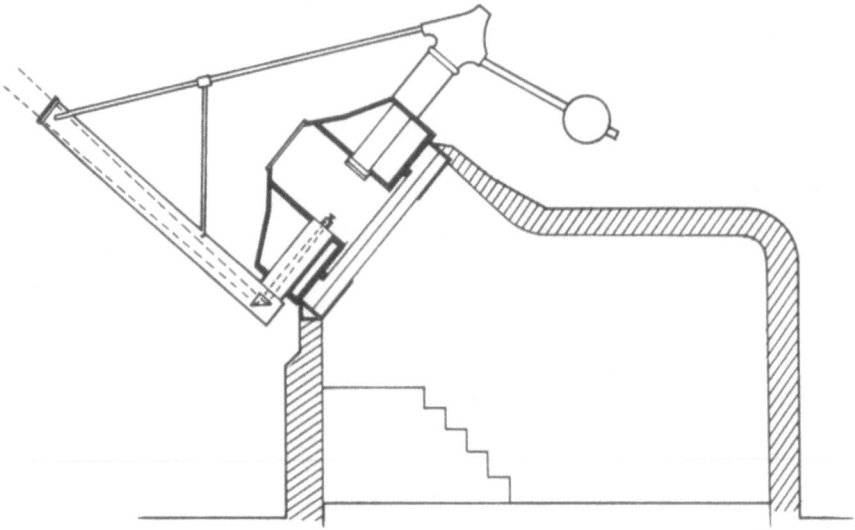

Fig. 10. Hartness turret telescope

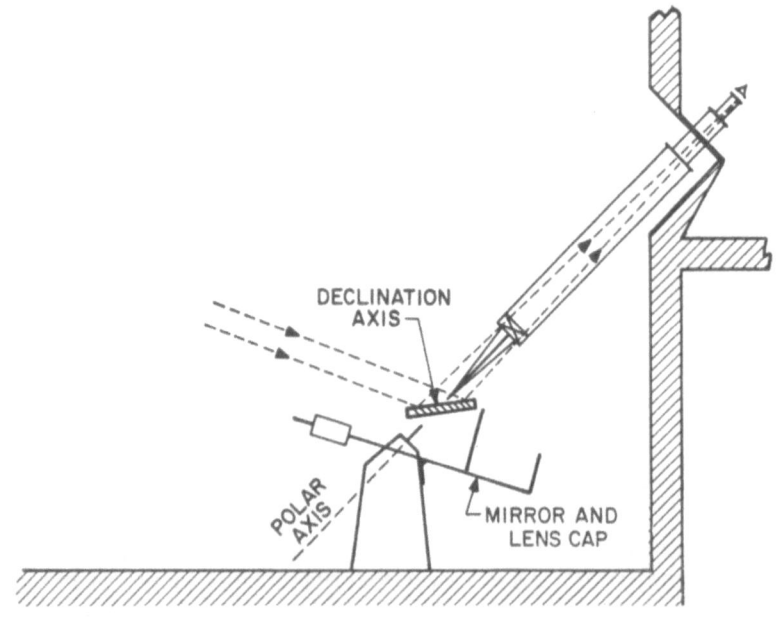

Fig. 11. Gerrish-type polar telescope

ment and advances in technology, a true parabola might indeed be achieved and one even with optical quality, provided that some presently very difficult features are ironed out (these include wrinkling of the material, asymmetric bending and cracking). Even hairline cracks are not acceptable. Another problem is to maintain the disc in its "parabolic mirror" form once it has been deformed. Various suggestions can be made, including the mechanical fastening of the sheet at a number of points, once the deformation has begun, to a substrate by epoxy or other plastic or even ceramo-plastic substances.

The above scheme unfortunately would not have too much merit for the Moon because we would still have to have the diameter of the desired mirror in the first place, and, as we have seen, the width is one of the problems in the packaging. Furthermore, the disc would have to be optically flat on one side which imposes the same sort of shipping and packaging problems as would be encountered with the optical paraboloid, if it were shipped in one piece.

We come then to our original suggestion that our astronomical lunar optics must take advantage of all advanced developments of light-weight optics and structures which are not only light, but very strong and easily erectable under very difficult conditions. A lunar telescope must not be restricted by the design and engineering notions that have been so successfully applied for over two hundred years to terrestrial telescopes. We should no more be fettered and bound

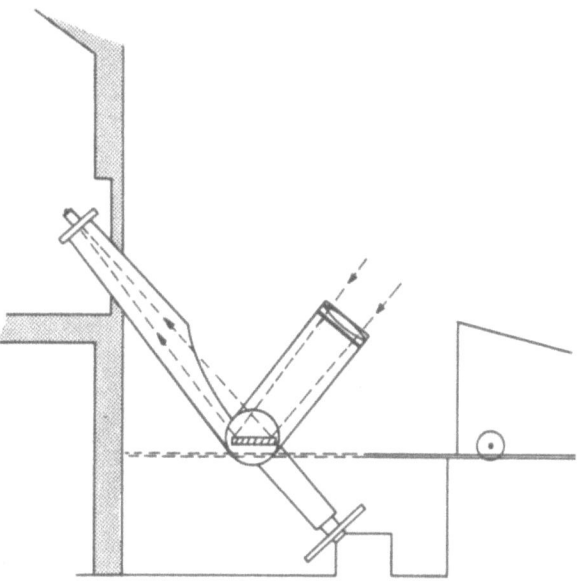

Fig. 12. Grubb modified Coudé telescope

by these terrestrially adequate notions than we would be for our rocket transportation, lunar landing system and our lunar surface transportation.

One example of an area that would need study and could use new design is the matter of housing. The astronomical dome and the heavy walls and piers so common with observatories which are usually placed on mountains serve very well to protect the instrument against the elements; not the astronomer, however! There are few places on Earth that seem colder than an observatory on a mountain in mid-winter! On the Moon there are no wind loadings or any weather environment, so such a structure would not be needed. If it is found that micrometeor activity on the Moon is neglibible, then such a telescope could very well stand on a prepared pad in the "open". This would very greatly ease the problems of mounting, steering, modifying and operating the equipment for the astronomer.

More than likely, it will be difficult, if not impossible, for the astronomer on the Moon to visually observe celestial objects because of the difficulty of getting near an eyepiece with a space suit, unless the entire machine were enclosed in pressurized housing — a most unlikely and probably useless encumberance. Therefore, photoelectric scanners and video imaging will be a necessity so that the astronomer, who wishes to work in a shirt sleeve environment, could retreat to his bunker or pressurized laboratory, and guide his instrument, and observe from the relative comfort of his laboratory. There is nothing at all impractical or impossible about the suggestion and enough photoelectric and video work has been done now, notably at the Dearborn Observatory, Northwestern University, to show that both "visual" scanning, astronomical body location and imaging of the desired target is entirely feasible with photovideo devices.

At the time we are ready to place a telescope on the Moon, remarkable improvements will have been achieved (based on our present knowledge of what is being planned and what is in the laboratory) to not only permit this remote control and operation of the telescope, but also to use the solid state or other quantum electronic devices to record the images for storage and analysis, or continuous visual inspection. A camera will still be used unless radiation damage to the plates is present. Again, we can use modern methods of xerography, and film transport changing to achieve the flexibility and resolution that are necessary. Photographic spectral and image recording is still the most acceptable, sensitive and practical scheme available to the astronomer. The astronomer need not work in the inimical environment of the lunar vacuum if the observatory can be housed and arranged as shown in Fig. 8—13.

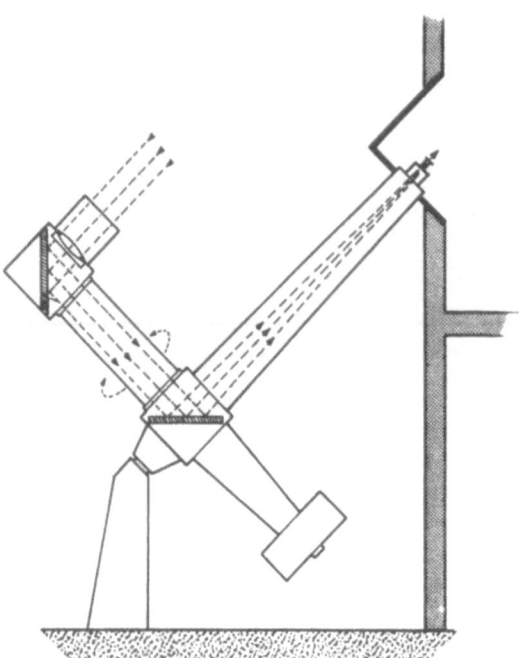

Fig. 13. Loewy's equatorial Coudé

We note that a variety of optical arrangements and trains has been suggested and tried in the past, for permitting the observer to sit in the relative comfort of an enclosed room or cabin, while guiding his telescope to any desired part of the heavens. In recent years the development of balloon astronomy by J. Strong, M. Schwarzschild, A. Hynek, A. Dollfus and others have shown the way for the development of smaller telescopes and spectrographs operated at high altitudes by remote control or by observer/pilots in a balloon cabin.

The "portable" LEM-based instruments shown in Figs. 15 and 16 could serve as initial pilot observatories. When and if a more permanent base were established, then larger optics could be installed in a more permanent fashion and protected, when necessary, from solar and other radiation and, perhaps, micrometeors. Such instruments, when arranged to permit visual observation and photography would powerfully complement photoelectric scanning, tracking and sensing.

The requirements for a radio observatory would be very much simpler since a guidable antenna and associated electronics would be needed. The major problem would be the transportation and erection of a light weight antenna whose aperture (if a concave "dish" were used) might be ten meters or more. The technology exists already for the fabrication and packaging of such antennas (Fig. 17).

Table III. *A Proposed Lunar Laboratory Establishment Sequence*

1. LEM-based and housed, small flexible catadioptric telescope — say 25 cm aperture.
2. a) Larger LEM-housed telescope/spectrograph — say 50 cm aperture.
   b) Small external radio-observatory antenna complex, with data center in LEM.
3. LEM-truck housed dual spectrograph/telescope with wide imaging flexibility.
4. a) Quasi-portable Lurain-based instruments and LEM-sheltered laboratory.
   b) Large Radio-observatory complex, with large antennas.
5. Lurain-based major instrument, say 150—400 cm aperture, buried in ground or otherwise protected and "permanent" underground laboratory.

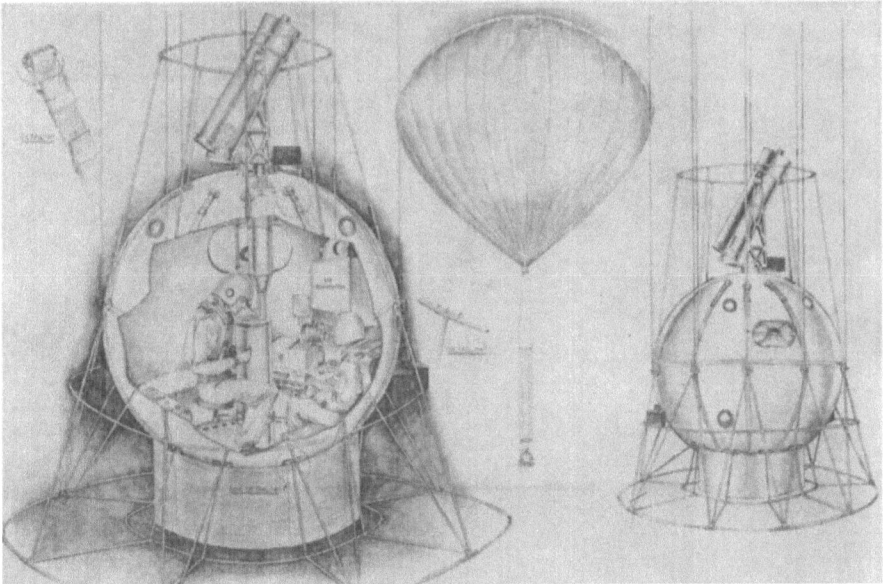

Fig. 14. Proposed use of the ONR-WRI Strato-Lab gondola for astronomical research flights. The Strato-Lab gondola is shown with a Schmidt telescope installed which can be oriented by independent rough and fine controls as well as a star-seeker by the observer from inside the pressurized capsule. Elevation and azimuth are controlled by electrical servo-mechanism. The gondola-balloon system is rigidly coupled and serves as the mass against which the telescope is rotated

Following this brief discussion of suggestions for some of the mechanical and optical arrangements for an observatory (see Table III), we list some of the elements of astronomy which could be fruitfully pursued on the lunar surface. At the outset, it is important to re-emphasize that we consider the observatory to be on the far side (from the Earth) of the Moon.

1. Astrometry

Advances in astrometry have recently disclosed the presence of large planets in orbits around some of the nearer stars. The detection of these objects and the tracing of their motions are so critical that skills and techniques of the highest magnitude are necessary to accumulate the data upon which these discoveries are based. PETER VAN DE KAMP has indicated that the principal advantage of the lunar observatory will be the absence of an atmosphere which will eliminate troublesome refraction. Refraction has always been a problem in positional astronomy for it alters the position of the stars and is even different for stars of different colors. Because of the change in position with time the differential refraction is proportional to exposure time which makes observations of faint stars susceptible to greater errors.

The absence of an atmosphere will also eliminate three important sources of systematic errors. Imperfect transparency, imperfect seeing and atmospheric

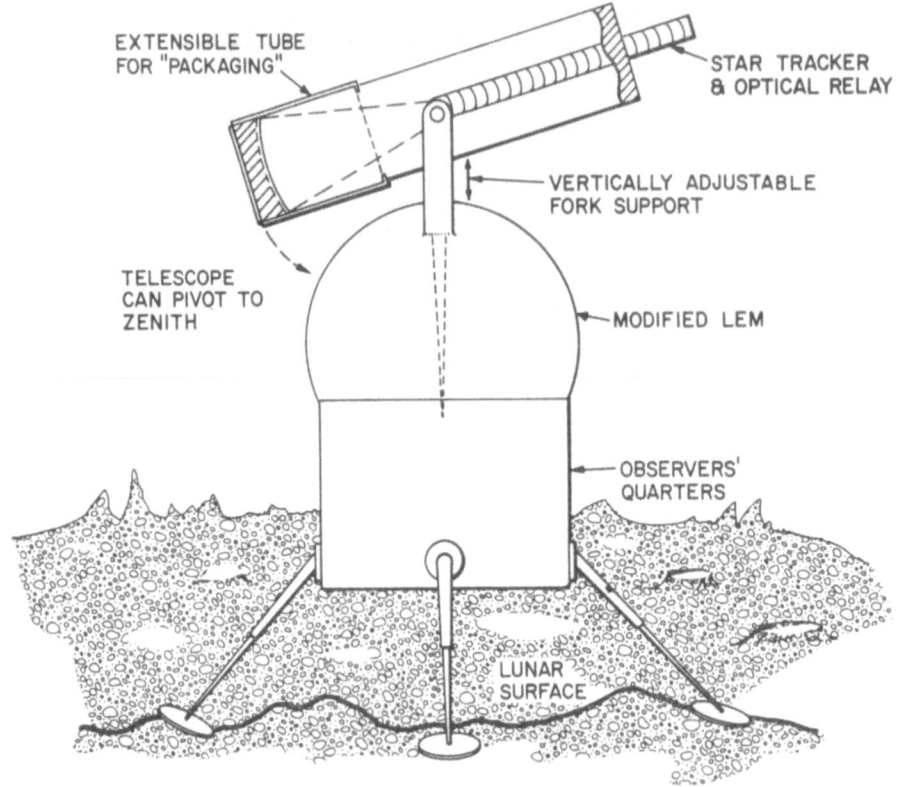

EXTENSIBLE TUBE
FOR "PACKAGING"

STAR TRACKER
& OPTICAL RELAY

VERTICALLY ADJUSTABLE
FORK SUPPORT

TELESCOPE
CAN PIVOT TO
ZENITH

MODIFIED LEM

OBSERVERS'
QUARTERS

LUNAR
SURFACE

Fig. 15. "LEM-based" catadioptric telescope

dispersion. The elimination of these errors will give rise to smaller stellar images on photographic plates, which means that fainter stars can be reached and more accurate measures made.

VAN DE KAMP indicates that there should be many stars of small mass with associated large planets in the vicinity of the Sun which have as yet not been detected. Actually, a star with a luminosity of about 1/100th that of the Sun can be detected, but when we go to fainter objects the critical factors involved in their detection preclude their discovery. Using mirror optics means that high optical resolution in the ultraviolet and infrared can be obtained to aid in the search for the faint, but highly important, white dwarf stars, and the ability to search in the infrared region may disclose the faint red dwarfs.

Another element in astrometrical observations could be the extention of parallax and proper motion studies to involve more distant stars. Perhaps one of the most important overtones from astrometric studies may be the overlapping of the various methods of determining the parallax of the stars, for it could precisely measure the parallaxes of stars as distant as 3,000 light years. This would provide more secure information on which to base our stellar distances.

L. SPITZER indicates that a 400-cm telescope could provide a resolution of 0.01 arc seconds. This could resolve a 50-foot object at 200,000 miles. But more important it could, for the first time, resolve surface features of a few of the supergiant stars like Alpha Orionis, Alpha Scorpii and Omicron Ceti whose diameters lie between 0.04 and 0.05 arc seconds.

2. Spectroscopy

There have been a great many model stellar atmospheres proposed to account for the observational evidence we find. By extending the spectrum into both the ultraviolet and infrared range verification may be obtained of one or more models

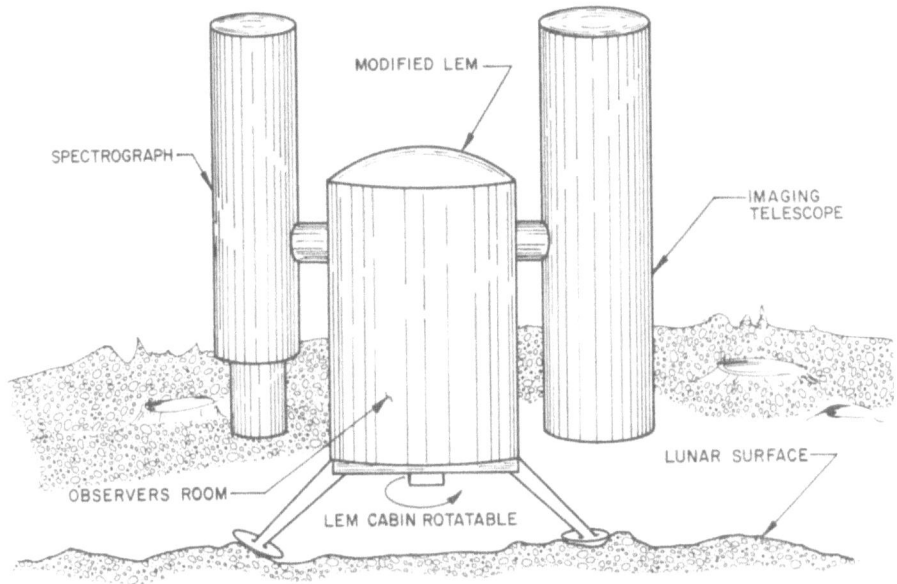

Fig. 16. Possible later version of "LEM-based" observatory

which corresponds to the observational evidence. We know that the cool stars will have a significant fraction of their total energy output concentrated in the infrared part of the spectrum. Likewise, a very hot star will have a significant portion of its energy concentrated in the ultraviolet part of the spectrum. Thus a lunar observatory can survey the sky in these portions of the electromagnetic spectrum, and in this fashion recognize the stars of various types and temperatures. It is highly likely that a survey of this type can provide many surprises, for some stars, perhaps too faint and too cool to be seen through the atmosphere, may show up in a survey on the Moon.

The lunar observatory would be an ideal place to study the gaseous nebulae for clues to the formation of the stars. In the Orion nebula photographic evidence indicates the formation of a small cluster of stars within the past two decades. However, had it been possible to scrutinize this region of space prior to 1947 the pre-birth details might have been revealed.

We know that scattered through the Milky Way are tremendous numbers of gaseous nebulae some as much as 100 light years in diameter. These are seen as systems whose components move and twist with speeds up to perhaps 10 miles a second. Knots and concentrations are also seen, and in the evolutionary processes taking place in these areas it is highly likely that the concentrations become protostars and eventually become stars.

A large telescope on the Moon would permit screening "suspicious" clouds with high magnifications to look for suspect areas where these protostars can form. Once

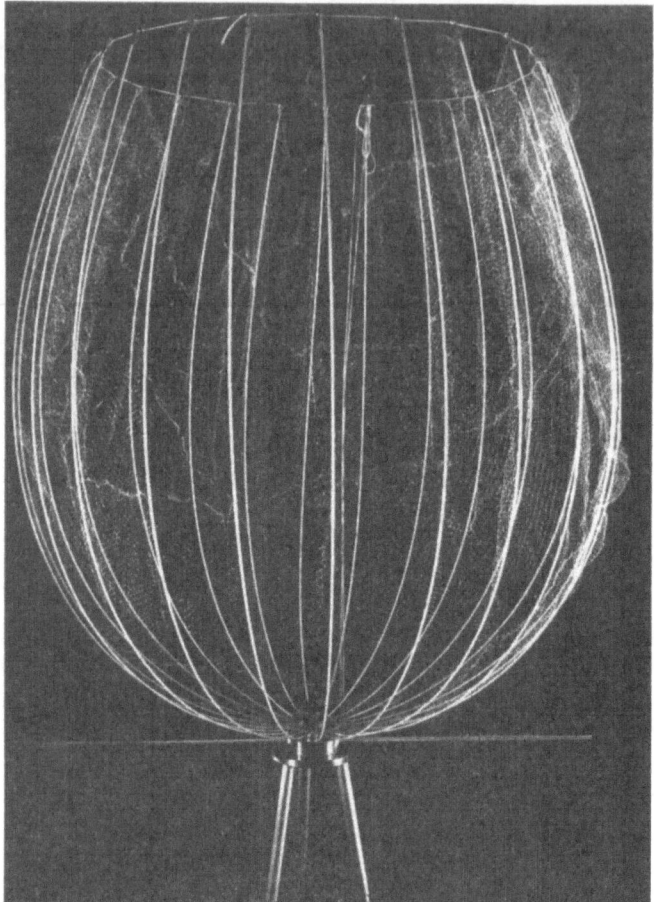

Fig. 17. Light weight radio astronomy antenna

these have been discovered then the spectroscope may be brought into play to screen these areas in the infrared for the actual detection of protostars. Some astronomers believe an infrared survey of the gaseous nebulae would provide the first clues to this event. Once there appears to be evidence that a star is "in the making" a concentrated observational program can be undertaken. The birth of a star marks one of the most important milestones in its life history. A lunar observatory may permit a viewing of this milestone.

The ultraviolet portions of the spectrum are important to the work of the astronomer. The fundamental lines of many of the elements such as carbon, nitrogen, oxygen and silicon are in the ultraviolet. The line profiles of the resonance lines for these elements are highly desirable.

Earth-based telescopes are able to observe only about 20% of the radiation from a BO star. This represents radiation on the red side of 3,000 Å. This means our knowledge of the total radiation of the O and early B stars is based on an extrapolation from the small fraction of their radiation in the visible part of the spectrum. Actual observations in the ultraviolet will improve our extrapolations, and may show up unexpected effects which may alter our ideas concerning the life histories of the stars. The total radiation in these stars determines the length of time the star can remain on the main sequence. It is also known that the distribution of light in the ultraviolet permits us to improve our measures of stellar temperatures which, with the total radiation, leads to an improved accuracy in our knowledge of the radii of the stars. The ultraviolet radiation distribution will reveal the opacity of the stellar atmosphere.

Spectroscopy practiced on the Moon could yield high resolution plates which could be interpreted by astronomers to yield temperatures, pressures, ionization, composition, magnetic fields and motions of the stars.

3. Sky Surveys

A wide angle Schmidt telescope would permit wide angle mapping of the sky in utralviolet and infrared. Other detectors may permit an X-ray and Gamma ray survey of the sky. This would permit a census of these objects to provide a better picture of the distribution of the various types of stars in the sky.

The use of a long focus telescope could photograph star clusters and these could be resolved because of the lack of bad seeing. The number of stars of various types in the clusters would help in the determination of the evolutionary trends in these objects.

4. Interstellar Physics and Galactic Studies

There are many unknowns concerning the formation of interstellar dust grains and their behavior. Similarly the interstellar radiation and magnetic fields require concentrated study to permit an understanding of the physical processes taking place in the galaxy. Even the content of interstellar space is only slightly understood, and this lack of knowledge affects the law of reddening of distant objects as we know them. In the galaxy are emission nebulosities of great extent whose relationship to the stars is only casually understood, and, finally, the non-thermal radiation of the interstellar medium can be studied. These represent special areas of interest in astronomy susceptible to resolution by the use of a lunar telescope.

A. B. MEINEL has indicated that if we look at a galaxy from outside the galaxy, and, if intergalactic space is relatively free of hydrogen atoms, then the outermost atmosphere of the galaxy will appear as a luminous envelope in Lyman alpha light. The reason for this is that all the radiations of shorter wavelength than 912 Å will be converted to this radiation by successive ionizations and radiative capture outward from individual stellar envelopes and circumstellar gas. While some of the radiation may be lost by absorption, the remaining radiation will be converted to Lyman alpha, and will escape from the galaxy. Thus, if there is no hydrogen in intergalactic space, and if the red shift is sufficiently large, the doppler shifted Lyman alpha will be detectable to provide information about the distant galaxies. The distant galaxies should be most interesting in Lyman alpha radiation.

If there is hydrogen in intergalactic space, then the absorption lines from distant galaxies should be sensitive indicators of the density of this gas. Because of the relative motions of our galaxy and the observed one, there should be a wide spread in the absorption line over a wide bandwidth, and this might provide an index to the distribution of hydrogen in the line of sight.

There are other cosmological problems susceptible to answer using a lunar observatory. With the large telescope galaxies can be counted to the limiting magnitudes. It is possible that we will be able to reach galaxies of magnitude 24 or 25 on the Moon and, thus, provide the astronomer with a sufficient number of the faint magnitudes. Galaxy diameters might be determined as a function of distance which would in turn yield information on world geometry.

The red shifts for high recessional velocities could be augmented by observations in the ultraviolet and infrared. By being able to use these regions we could provide a hold on many more lines than now available. This would give a better hold on such quasars as 3C-9, where the red shifted Lyman alpha line is tied to only one other line (C 4), and the 3C-286, where only one line is visible.

The most significant contribution made by a lunar observatory can well be the resolution of the problem of cosmological models. This is perhaps of even greater significance than the increase in distance which will extend our observations to a larger fraction of the universe. Astronomers believe that we may be able to reach the region where a choice can be made of the various models which have been proposed.

With the clasical work of Hubble on the extragalactic objects, the edge of the universe has been constantly extended, and simultaneously we have been indoctrinated in the expanding or evolutionary universe concept. In this model the universe began at some 12 to 15 billion years ago, and is still expanding, so that all galaxies are receding from each other with speeds proportional to their distances. Shortly thereafter it was suspected that another solution to the mathematics involved permitted a pulsating universe — one which expanded for a time, and then concentrated back to the same primeval nucleus. The period of this oscillation is long, and may reach many tens of billions of years. Of rather recent origin is the steady-state universe proposed by British cosmologists. These are the various models from which one may choose.

Hoyle has advanced some rather cogent arguments which favor his steady-state theory. He indicates that since this model requires no difference in large-scale properties between the past and the present, the theory is clearly susceptible to checking by penetrating deeply into space. He indicated that large-scale properties can be estimated from many different clues. These could be the populations of galaxies, the magnitude and color of their light, the radio emissions signaling collisions and other significant events, the relation between the red shift and the distance of the galaxies, etc.

He reasons that in the steady-state universe we should expect to see surviving only properties which have stabilized themselves, so that they are reproduced at precisely the same level from generation to generation. Galaxies represent a strictly controlled system with the origin of matter cast in a critical role. The crucial difference between the models can form the basis of stringent tests. The tests can be applied to such properties as the density of galaxies in space and the distribution of sizes and masses of galaxies. It is possible to check to determine whether distribution follows a regular frequency curve or shows no regular pattern.

Cosmologists who favor the expanding universe concept indicate that by penetrating deeply into space we are also moving back in time. If we live in this evolutionary universe, and galaxies do change slightly with age, then a change in form, in light, in substance of only a few percent in a billion years can readily be determined, if we can penetrate far enough into space to go far enough back into time. If such a change should be indicated it would be indicative of an evolutionary universe. It is this feature of deeper penetration in space that might be provided by a lunar observatory.

One other overtone related to the problem of cosmological models involves the use of radio telescopes and will be discussed in the next section.

5. Radio Astronomy

By getting away from the Earth with its ionospheric cutoff, the resolution of a radio telescope will be limited only by the achievable size of the directive antenna at long wavelengths. As the optical observatory site will be on the hidden side of the Moon, the accompanying radio telescope will also be placed here. It is this site which will permit the exploration of the radio spectrum in all wavelengths.

There exists today a map of radio sources at 18.3 megacycles over a portion of the celestial sphere. The reason for this wavelength is the inability of radio signals of higher frequency to penetrate the ionosphere. One of the first tasks would be to complete a map of radio sources in the Milky Way over a wide spectrum of frequencies.

Ionized hydrogen because of its absorption characteristics is important in determining the brightness distribution on the celestial sphere. In the low frequency regime, that is, below one or two megacycles this absorption exists, and it is anticipated that observations of absorption near the galactic plane may reveal clouds heretofore unsuspected.

One of the more fascinating studies today is associated with the radio emissions coming from the planets. A radio telescope on the Moon would be able to study these in many wavelengths. This along with spectroscopic studies of the atmospheres, especially of Mars and Venus, may resolve many of the problems associated with these bodies.

Radio astronomy offers the possibility of something close to a direct test of the creation of matter in space to verify the steady-state concept. The total amount of matter in the galaxy, if uniformly spread across the galaxy, is estimated as about 10^{-30} g/cc. The steady-state theory predicts that the average density of matter should be 10 times or more greater. The difference is accounted for by hydrogen spread through intergalactic space. Up to now there has been no possibility of detecting intergalactic matter. But a radio telescope on the Moon tuned to the 21 cm wavelength may be able to discover quantiatively the hydrogen distribution that may exist in intergalactic space.

Acknowledgements

The senior author acknowledges with thanks the suggestions on light-weight optics given by Messrs. C. STEPHENS and R. TEMPLE of Electro-Optical Systems, Inc., and various helpful comments by Drs. J. KUPPERIAN and E. STUHLINGER, NASA.

References

1. R. W. WOOD, Amateur Telescope Making. Scientific American, p. 323 (1933).
2. K. R. STEHLING, Operating on the Moon, Part II. Space/Aeronautics Magazine, p. 42 (February 1961).
3. I. M. LEVITT, Target for Tomorrow. New York: Fleet and Co., 1959.
4. M. FAGET, Apollo — The Long View. Astronautics and Aeronautics, p. 61 (April, 1963).

5. L. Spitzer, Jr., The Beginnings and Future of Space Astronomy. American Scientist **50**, No. 3 (1962).
6. A. B. Meinel, Astronomical Observations from Space Vehicles. Astronomical Society of Pacific Journ. **71**, No. 422 (1959).
7. P. van de Kamp, The Discovery of Planetary Companions of Stars. Yale Scientific Magazine **38**, 6 (1963).
8. M. Schwarzshild, J. B. Rogerson, Jr., and J. W. Evans, Solar Photographs from 80,000 Feet. Astronomical Journ. **63**, 313 (1958).
9. F. J. Malina, Lunar International Laboratory. Space Flight **7**, 155 (1965).
10. F. J. Malina, Report of the Lunar International Laboratory Discussion Panel, Warsaw. Astronaut. Acta **11**, 123 (1965).

Astronomical Telescope on the Moon

By

V. M. Mozjherin[1], V. B. Nikonov[1], V. K. Prokofiev[1] and N. S. Chernykh[1]

(With 2 Figures)

Abstract — Résumé — Резюме

Astronomical Telescope on the Moon. It is desirable in connection with a Lunar International Laboratory to establish on the Moon an astronomical telescope which should either be fixed in relation to the Moon or else shifted at specified intervals according to the programme being implemented or the requirements of the observer. Owing to the physical movements of the Moon (libration, precession) it is possible in the course of a year, with a telescope fixed in relation to the Moon's surface, to observe a certain belt of the star field. The axis of the telescope will describe a complicated trajectory of the star field. The setting of the telescope can subsequently be adjusted so as to cover the most interesting parts of the star field, the scanning of which will then be effected as a result of the natural rotation of the Moon.

It is extremely important to obtain television pictures of the galactic fields with different filters. For this purpose, the axis of the telescope is trained onto the appropriate zone of the sky; and its field of vision is traversed by a series of galaxies as a result of the rotation of the Moon itself. When the field of vision of the telescope is 15′, each galaxy can be observed on the screen for a period of 30 minutes.

The radiation of the planets can be registered practically through the full orbit, right up to positions very close to the Sun; the Sun's disc, seen from the Moon, has no halo. In particular, it is possible to observe Venus in inferior conjunction with the Sun: in this case it is possible to follow the absorption of the Sun's light in the upper strata of the atmosphere of Venus in different parts of the spectrum, up to and including very short wave lengths. This will provide extremely valuable data for ascertaining the structure and composition of the upper strata of the atmosphere of Venus.

Télescope astronomique sur la Lune. Dans le cadre d'un Laboratoire Lunaire International, il faut considérer qu'il est utile d'installer sur la lune un télescope astronomique soit immobile par rapport à la lune soit avec changement de position a des intervalles de temps déterminés par un programme ou par un observateur. Les mouvements physiques de la lune (libration, precession) permettent d'observer au, cours d'une année, une ceinture déterminée du ciel étoilé à l'aide du télescope immobile à la surface de la lune. L'axe du télescope décrira sur le ciel étoilé une trajectoire complexe. A la mise en place du télescope, on peut orienter judicieusement le secteur de cette trajectoire vers les régions étoilées les plus intéressantes; l'exploration de ces régions connues par le calcul résultera de la rotation propre de la lune.

Il parait très important d'obtenir des images télévisées des champs de galaxies avec différents filtres. La trajectoire de l'axe du télescope sera dirigée pour cela sur la région utile; par la rotation propre de la lune, une série de galaxies, connue par le calcul, viendra dans le champ de vision du télescope; avec un champ de vision du

---

[1] Crimean Observatory, Nauchny, Crimea, U.S.S.R.

télescope de l'ordre de 15′ une galaxie isolée peut apparaître sur l'écran pendant 30 minutes.

L'enregistrement du rayonnement des planètes est possible sur l'orbite presque complète, jusqu'aux positions très proches du soleil; le disque solaire, tel qu'on le voit de la lune, n'a pas de halo. On peut, en particulier, observer Vénus très près de la conjonction inférieure, on peut alors suivre l'absorption de la lumière solaire dans les couches supérieures de l'atmosphère de Vénus, dans différentes parties du spectre, jusqu'aux très courtes longueurs d'ondes. Ceci apporte des renseignements très précieux permettant de se faire une opinion sur la structure et la composition des couches supérieures de l'atmosphère de Vénus.

Астрономический телескоп на Луне. Следует считать целесообразным установку на Луне в составе Международной Лунной лаборатории астрономического телескопа либо неподвижно относительно Луны, либо с изменением положения через определенные промежутки времени по программе или наблюдателем. Физические движения Луны (физическая либрация, прецессия) позволяют осмотреть неподвижным относительно Лунной поверхности телескопом за год некоторый пояс звездного неба. Ось телескопа будет описывать на звездном небе сложную траекторию. Установкой телескопа можно участок этой траектории направлять последовательно, на наиболее интересные звездные области; сканирование этих областей будет происходить за счет естественного вращения Луны.

Представляется очень важным получение телевизионных изображений полей галактик с разными фильтрами. Траектория оси телескопа направляется при этом на нужную область; за счет собственного вращения Луны в поле зрения телескопа пройдет ряд галактик; при поле зрения телескопа 15′ отдельную галактику можно наблюдать на экране в течение 30 минут времени.

Регистрация излучения планет возможна почти на полной орбите вплоть до очень близких положений около Солнца; диск Солнца, видимый с Луны, не имеет ореола. В частности возможно наблюдение Венеры вблизи нижнего соединения, когда можно проследить поглощение солнечного света в верхних слоях атмосферы Венеры в разных участках спектра, вплоть до очень коротких длин волн. Это даст весьма ценные сведения для суждения о структуре и составе верхних слоев атмосферы Венеры.

We consider that a Lunar International Laboratory should, as a matter of course, include an astronomical telescope for carrying out specific astronomical observations. We shall not deal here with the technical problems of installing the telescope on the Moon; neither shall we touch on either its technical parameters or the conditions, more especially the temperature conditions, in which it functions. All these are subjects requiring a special study. We propose, in this article, to consider the best methods of carrying out observations with the telescope, making use of the special features of the Moon's movement in its orbit and rotation round its own axis.

In the event of the existence on the Moon of a manned laboratory, the telescope could be used by the laboratory staff for carrying out observations; but it would, we think, be preferable for it to be operated automatically. In that case, researchers on the Moon would be able to set the telescope a programme of work for a specific period of time; so that they themselves could concentrate on other tasks. Since the laboratory staff will in all probability be small in relation to the vast amount of work to be done, time-saving will be an important factor in organizing the functioning of a Lunar Laboratory. Moreover, it is essential to allow for the fact that, once a laboratory begins operating, its staff will probably be faced with various new problems not covered by the programme, such as repairs, observation of new phenomena, and so on. In view of all this, we have come to the conclusion that the telescope should be designed, mainly, to operate automatically.

Let us assume that the telescope is installed permanently at a point A on the Moon's surface in the plane of the local meridian at an angle ϱ to the axis of rotation of the Moon (Fig. 1 gives a section in the plane of the local meridian). At the point B, with selenographic latitude $\varphi = 90 - \varrho$, the axis of the telescope will be trained on the zenith Z'. We shall determine the trajectory described by the axis of the telescope as a result of the Moon's rotation about its own axis; and we shall determine the coordinates of the points on this trajectory in the terrestrial equatorial coordinates system.

The pole of the Lunar equator is divergent from the pole of the ecliptic by 1.5°; thus for approximate bearing on the Moon, star charts in the ecliptic coordinate

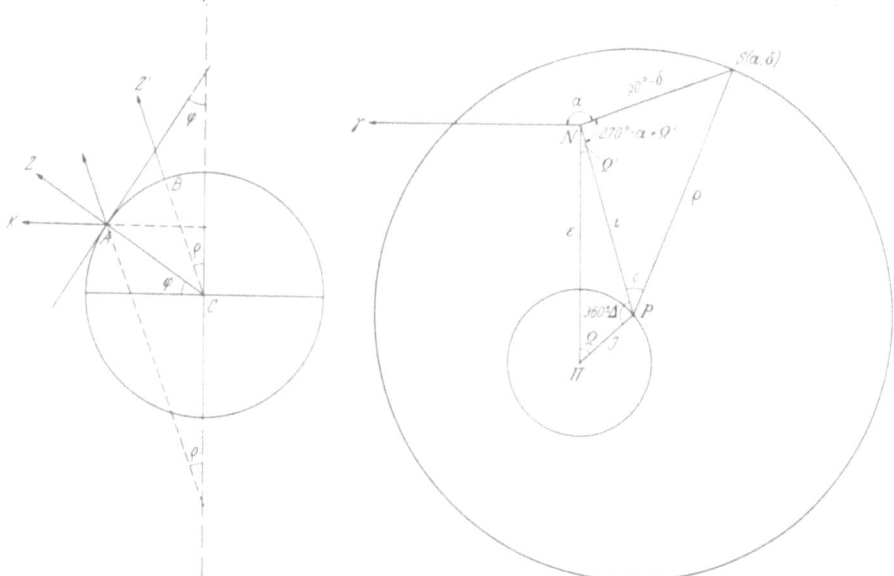

Fig. 1. Section of the Moon in the plane of the local meridian

Fig. 2. Part of the geocentric celestial sphere, observed from within

system are perfectly adequate. For more accurate bearings, the catalogue of star positions in the Lunar coordinate system can be used [1]. On the basis of a fairly detailed catalogue of the mean star positions for every 10—20 days, it will be easy to calculate which stars will fall within the field of vision of a fixed telescope as a result of the daily rotation of the Moon.

We shall describe the method for calculating the coordinates of the points on the curve described by the optical axis of the lunar telescope, in the terrestrial equatorial system. In the interest of simplicity, we shall consider only the precessional movement of the pole of the lunar equator, ignoring both mutation (physical libration) and aberration. The resulting error in the required coordinates may be of the order of 2—3 minutes of the arc, which is fully acceptable with a telescope field of vision of 15'—20'.

Fig. 2 shows part of the geocentric celestial sphere, observed from within: Π — north pole of the ecliptic, N — north pole of the terrestrial equator, P — north pole of the mean lunar equator, S — path of the optical axis of the telescope on the celestial sphere, γ — vernal equinox.

In the triangle $N\Pi P$ all the factors are known: ε and J may be considered invariable; i, Ω', $\varDelta$ and Ω occur in the Astronomical Yearbooks at intervals of 10 days.

As already stated, the telescope will be installed on the meridian at a point A

on the Moon at an angle ϱ to the Moon's rotation axis (Fig. 1). Thus the seleno-centric polar distance ϱ of the optical axis of the telescope is given; the angle ϱ is known and, with a stable position of the telescope, remains constant. Due to the daily rotation of the Moon, the path S of the optical axis of the telescope, over a period of 27.3 days, will describe a small circle of radius ϱ about P. As a result of precession, the position of the point P will shift constantly, so that the path S of the optical axis will describe, on the celestial sphere, a continuous spiral-shaped curve. The terrestrial equatorial coordinates α and δ of the points on this curve may be found by examination of the spherical triangle NPS. In this triangle two sides, i and ϱ, are known. Taking a certain value α, we shall find, with this, the angle $PNS = 270° - \alpha + \Omega'$.

Resolution of the spherical triangle PNS will make it possible to find the declination δ. We first calculate the auxiliary angle θ,

$$\text{tg } \theta = \cos i \, \text{tg } (270° - \alpha + \Omega') \tag{1}$$

The angle $C = SPN$ is then determined by the formula

$$\sin(C + \theta) = \frac{\text{tg } i \sin \theta}{\text{tg } \varrho} \tag{2}$$

And finally, the required value of the declination δ

$$\cos \delta = \frac{\sin \varrho \sin C}{\sin(270° - \alpha + \Omega')} \tag{3}$$

Repeating this process for a whole series of fixed values of α, we obtain the corresponding values of δ for the points on the curve with which we are concerned. We give in Table I, as an example, the values thus obtained of the terrestrial equatorial

Table I

α		δ		α		δ
1968 January 0.0	6h	67°01'	1968 January 0.0	20h	27°10'
	7	66 34		21	31 46
	8	65 11		22	37 38
	9	62 51		23	44 04
	10	59 24		00	50 21
	11	54 45		1	55 52
	12	49 03		2	60 14
	13	42 39		3	63 25
	14	36 15		4	65 32
	15	30 38		5	66 42
	16	26 20	January 27.3	6	67 01
	17	23 49		7	66 33
	18	22 57		8	65 11
	19	24 05			

coordinates of the trajectory of the centre of the telescope's field of vision, installed at an angle $\varrho = 45°$, throughout one lunar rotation in the period 1.0 to 28.5 January 1968, at one-hour intervals for right ascension.

Analysis of the formulae given shows that, with $\varrho > i$, the right ascensions α of points on the curve may have any value from 0^h to 24^h. For each value of α these formulae give one value of δ.

With $\varrho \leq i$, the curve lies in the limited region of right ascensions from α' to α'', determined by the equations:

$$\left. \begin{array}{l} \cos(\alpha'' - \Omega') = -\dfrac{\sin\varrho}{\sin i} \\[3mm] \cos(\alpha' - \Omega') = \dfrac{\sin\varrho}{\sin i} \end{array} \right\} \qquad (4)$$

To each value of α, in this case, there correspond two values of δ. Note also that when, according to the formula (3) we obtain $\cos\delta < 0$, we have to take $\delta < 0$.

The calculation of the coordinates of the points on the curve for one or two rotations of the Moon may be made with constant values of Ω' and i. In order, however, to trace the movement of the optical axis of the telescope on the celestial sphere for a lengthy period (more than one or two sidereal months) the calculations have to be made with Ω' and i values taken at moments of time close to the moment of upper culmination at a particular point on the Moon viewed from that point of the curve. It may be reckoned with sufficient accuracy for $\varrho > 10°$ that, on a point on the Moon with selenographic longitude l in the upper culmination there will be found, at a moment t of universal time, stars with longitude

$$\lambda_\mu = \mathbb{C} \pm 12^h + l - l_0 \qquad (5)$$

where $\mathbb{C}$ is the mean longitude of the Moon at that moment, and l_0 is the libration of the moon in longitude.

The right ascensions of these stars may be found, approximately, by means of the formula:

$$\text{tg } \alpha_\mu = \frac{0.92 \sin\lambda_\mu - 0.40 \text{ ctg } \varrho}{\cos\lambda_\mu} \qquad (6)$$

obtained from the formulae for transformation of elliptical coordinates into equatorial ones by substitution $\beta \approx 90 - \varrho$.

In regions close to the pole of the terrestrial equator, the error of the formula (6) will attain tens of degrees, but even in this case the value α_μ will be entirely adequate for purposes of calculation.

Owing to the precession movement of the pole P, each successive spiral of the curve described by the optical axis of the telescope on the celestial sphere swings round the pole of the ecliptic Π at an angle $1°26'48''.4$, the declination of the points on one half of this curve increasing, and on the other half, decreasing. The greatest difference in declination, attaining $2'21''.7$ in the course of a sidereal month, occurs for those points on the curve whose right ascensions are determined approximately by the equations:

$$\left. \begin{array}{l} \text{ctg } (270° - \alpha_1 + \Omega') = \dfrac{\cos\varrho \sin i + \sin\varrho \cos i \sin\Omega}{\sin\varrho \cos\Omega} \\[3mm] \text{ctg } (\alpha_2 - \Omega' - 270°) = \dfrac{\cos\varrho \sin i - \sin\varrho \cos i \sin\Omega}{\sin\varrho \cos\Omega} \end{array} \right\} \qquad (7)$$

At points equidistant from the points with right ascension α_1 and α_2, the differences of declination are zero. The rate of change of declination may be taken as constant over a period of several sidereal months. With the maximum rate of change of declination, $2'.3$ in a sidereal month, the displacement of a star located in the region close to α_1 and α_2, in the case of a telescope with $15'$ field of vision, will take 6 sidereal months — in other words, a star located in this region can be observed 6 times. For stars having intermediate right ascensions the rate of change of declination will be slower, and it will be possible to observe then in the course of a greater number of sidereal revolutions, up to a maximum of 40.

It is clear from the above description of the movement of a telescopic axis that, in order to observe a particular section of the celestial vault, the telescope must be set at the corresponding angle ϱ. For example, with $\varrho \approx 90°$ (AK in Fig. 1), the telescope will observe the region of the ecliptic, and register the planets located within this region. There will obviously be, according to the selenographic latitude φ at which the telescope is set up, some parts of the star sky which cannot be observed. The region which escapes observation would be that lying within a cone with an angle 2φ at the apex. Obviously it is necessary to account for the effect of precession. Only when the telescope is set up at a point close to the Lunar equator will it be possible to observe the whole of the celestial vault. It will therefore be expedient to install the telescope in the vicinity of the Lunar equator.

It may sometimes be necessary to register a specific object in the sky. In that case the angle ϱ for setting the telescope can be found from the spherical triangle PNS on the basis of the α and δ values of this object, which are given. The setting of the telescope on to the target will be effected in two stages, as follows: firstly, the calculated angle ϱ of inclination of the telescope in the plane of the local meridian will be set either automatically or else by the observer; subsequently, we shall have to wait until, with the rotation of the Moon about its own axis, the required target attains its culmination. It would be possible to construct the telescope in such a way as to allow for an azimuth setting, but this would be more complicated, besides which it is not necessary. There are many targets to observe, and a programme can be drawn up for each revolution of the Moon, such that each target can be observed in turn as it attains its culmination. In this way, also, time lost in waiting can be substantially reduced.

The Moon's rotation about its axis is fairly slow, so that the axis of the telescope traverses the celestial vault at a rate of 30″ per minute. Thus, a telescope with a field of vision of 15′ will observe a target for the duration of half an hour. If the telescope is equipped with a television camera, it will be possible, during this time, to obtain several pictures of a galaxy in different parts of a wide region of the spectrum, from infrared to short-waves (to approximately 1,000 Å).

Coaxially with the telescope, it will be useful to set up an X-ray telescope as well. The combination of the two will then provide sufficiently complete information about the short-wave radiation of galaxies and other celestial targets.

The planning of the telescope's programme of work, based on the angles ϱ, may be done by an observer stationed either on the Moon or on the Earth. In the latter case, a terrestrial point from which the Moon is within the field of direct visibility, transmits the ϱ angle settings corresponding to the times when observations are to be carried out.

With this system, the degree to which the telescope can be controlled is reduced; but it has the advantage of simplifying the process for the setting of one angle only.

As regards the apparatus with which the telescope should be fitted, we have already mentioned a television camera for transmitting to Earth pictures of the galaxies. For spectral observations, it will be essential to have a small multichannel spectrograph for making spectrophotometric measurements of stars in relatively wide sections of the spectrum ($\Delta\lambda \approx 200-500$ Å), i.e., for multicolour photometry of stars at wave-lengths of between 1,000 Å and a few microns. The television apparatus can of course also be used for multicolour photometry of the stars; for this purpose, it will have to transmit pictures of the section of a star field traversing the field of vision of the telescope, using for this purpose a number of different filters (as for pictures of the galaxies).

There are many possible variations of equipment to be used with this telescope, but we do not intend to describe them in detail here. We should merely like to state, in conclusion, that a relatively small telescope, with a 500 mm mirror and equivalent focal length of 6 metres, is capable of effecting multicolour photometric measurements of stars up to a star magnitude of 12−13.

References

1. A. A. YAKOVKIN, I. M. DEMENKO and L. N. MIZ, Formulae and Ephemerides for Field Observations on the Moon. Kiev, 1964.
2. The Astronomical Yearbook of the U.S.S.R. for 1968. Moscow, 1964.

Lunar Radio Astronomy Observatory

By

Stanislaw Gorgolewski[1]

Abstract — Résumé — Резюме

Lunar Radio Astronomy Observatory. Terrestrial, orbital and lunar environments are compared from the radio-astronomical point of view. The terrestrial environment is limited at long waves by the ionosphere, atmospheric noise and man-made interference. The centimeter and shorter waves are plagued by atmospheric absorption and radiation. Earth orbiting radio telescopes may considerably reduce these effects, yet, even then, the ionospheric and exospheric noise is important at long waves, and man-made interference will still be present almost throughout the whole radio spectrum. The lunar environment on the opposite side of the Moon should be almost completely free from the above mentioned limitations.

The instruments proposed for the Lunar Radio Astronomy Observatory include interferometers for aperture synthesis in the kilometer, hectometer and decameter wave ranges. These include aerials based on the Moon or orbiting the Moon. Laser or radio links could be used for information transfer to the receivers. The use of lunar craters for stationary radio telescopes is considered as well as the employment of inflated structures using polymerizing plastic filling for steerable aerial types. Accurate dishes are needed for decimeter, centimeter and millimeter wave ranges with cooled parametric amplifiers, masers or thermal detection radiometers in the focus. Aperture synthesis is again recommended in the decimeter and even shorter waves.

Some possible types of research at a Lunar Radio Astronomy Observatory during lunar nighttime include the following: galactic and extra-galactic low angular resolution spectral studies at very low frequencies concerned with the bulk matter distribution with increasing "redshift". Very important are high resolution observations of low frequency galactic radiations and of a large number of extra-galactic radio sources. The very low frequency spectra of extra-galactic radio-sources should show progressive shifts of their frequencies of maximal flux densities toward longer waves with increasing "red shift". Thus it should be possible to measure their "red shifts" by radio observations alone. The lunar environment is also advantageous for the 21 cm hydrogen line and OH line studies. It should also help us in the achievement of ultimate sensitivities in the decimeter and centimeter wave ranges. The stronger radio sources, the galactic center, the planets, H_{II} regions and some galactic nebulae are among the possible targets in this spectral range.

The lunar day will be unfavourable for all high sensitivity observations, because the Sun is the dominant radio source in the sky, especially at shorter waves. The main object for studies at this time will be the Sun itself and the solar corona.

Observatoire Radio-Astronomique Lunaire. On compare, du point de vue de la radioastronomie, les conditions d'ambiance terrestres, orbitales et lunaires. L'environnement terrestre est limité, dans la gamme des ondes longues, par l'ionosphère, le bruit atmosphérique et les parasites produits par l'Homme. Les ondes centimétriques

---

[1] Astronomical Observatory, N. Copernicus University, Toruń, Poland.

et plus courtes sont fortement réduites par l'absorption et le rayonnement atmosphérique. Des radiotélescopes orbitant autour de la terre peuvent diminuer considérablement ces effets, mais le bruit ionosphérique et endosphérique continuera d'être important en grandes ondes, et les parasites produits par l'homme seront toujours présents sur toute l'étendue du spectre radioélectrique. Le site lunaire, sur la face opposée de la Lune, devrait presque être complètement exempt de ces inconvénients.

Les instruments proposés pour un observatoire radio-astronomique lunaire comprennent des interféromètres pour la synthèse d'ouverture dans les bandes kilométriques, hectométriques et décamétriques. Ces instruments comprennent des antennes posées sur la Lune, ou orbitant autour d'elle. Des liaisons laser ou radio pourraient être utilisées pour transmettre l'information vers les récepteurs. Pour l'installation de radiotélescopes fixes, on prévoit aussi bien l'utilisation de cratères lunaires, que celle d'enveloppes en plastique polymère pour les antennes orientables. Des paraboles très précises sont nécessaires dans les bandes décimétrique, centimétrique et millimétrique; on placera à leur foyer un amplificateur paramétrique avec refroidissement, un maser ou un radiomètre détecteur de rayonnement thermique. La synthèse d'ouverture est encore recommandée dans la bande décimétrique et plus courte.

Parmi les recherches possibles à partir de l'Observatoire radio-astronomique lunaire au cours de la nuit lunaire on trouve: les études spectrales galactiques et extra-galactiques à faible résolution angulaire aux très basses fréquences portant sur les distributions d'amas galactiques présentant un décalage croissant vers le rouge. Les observations, avec une grande résolution, de rayonnements galactiques en basse fréquence et d'un très grand nombre de radio-sources extra-galactiques sont très importantes. Les spectres en très basse fréquence des radio-sources extra-galactiques devraient présenter des décalages progressifs de leurs fréquences de densités de flux maximaux vers des ondes plus longues avec un glissement croissant vers le rouge. De ce fait, ce glissement pourrait être mesuré par les seules observations radio. Le site lunaire est également très avantageux pour l'étude des raies d'hydrogène en 21 cm et de OH. Il devrait également permettre d'atteindre les sensibilités limites dans les bandes décimétrique et centimétrique. Les cibles possibles dans cette bande sont: les radio-sources plus intenses, le centre de la galaxie, les planètes, les régions H$_{II}$ et certaines nébuleuses galactiques.

Le jour lunaire sera défavorable à toutes les observations à haute sensibilité, du fait que le Soleil constitue la radio-source dominante du ciel, particulièrement aux ondes les plus courtes. Dans le présent, l'objet principal des études reste le Soleil lui-même, ainsi que la couronne solaire.

Лунная радиоастрономическая обсерватория. Земная, орбитальная и лунная среда сравниваются с радиоастрономической точки зрения. Земная среда ограничивается в диапазоне длинных волн ионосферой, атмосферными шумами и искусственными помехами. Сантиметровые и более короткие волны подвергаются отрицательному воздействию атмосферного поглощения и излучения. Выведение радиотелескопов на земную орбиту может значительно снизить воздействие этих помех, однако даже в этом случае ионосферные и экзосферные шумы в значительной степени ощущаются на длинных волнах, а искусственные помехи будут сохраняться почти на всем диапазоне спектра радиоволн. Лунная среда на обратной стороне луны должна быть почти совершенно свободна от вышеуказанных помех.

Предлагаемая аппаратура для лунной радиоастрономической обсерватории включает интерферометры для апертурного синтеза в диапазонах километровых, гектометровых и декаметровых волн. Сюда входят антенны, помещенные на луне или выведенные на орбиту вокруг луны. Для передачи информации в приемники могут быть использованы лазеры или линии радиосвязи. Рассматривается возможность использования лунных кратеров для стационарных радиотелескопов, а также применения заполненных структур с использованием наполнения из полимеров для антенн с управляемой диаграммой. Для дециметровых, сантиметровых и миллиметровых волн требуются точные чашеобразные антенны с охлаждаемыми параметрическими усилителями, мазерами или теплоулавливающими радиометрами в фокусе. Апертурный синтез также рекомендуется в диапазоне дециметровых и даже более коротких волн.

В число возможных типов исследований на лунной радиоастрономической обсерватории в период лунной ночи входят следующие: галактические и внегалактические спектральные исследовании при низкой угловой разрешающей способности на очень низких частотах, касающиеся распределения массы материи при усиливающемся „красном смещении". Очень важны наблюдения при высокой разрешающей способности по угловым координатам низкочастотных галактических радиаций и большого числа внегалактических источников радиоизлучения. Самые низкочастотные спектры внегалактических источников радиоизлучения должны показывать постепенные смещения их частот максимальной плотности потока в направлении более длинных волн при усиливающемся „красном смещении". Таким образом можно будет измерить их „красное смещение" лишь с помощью радионаблюдений. Лунная среда также благоприятна для изучения 21 см линии водорода и линии OH. Она также должна способствовать достижению предельной чувствительности в диапазонах дециметровых и сантиметровых волн. Более сильные источники радиоизлучения, центр галактики, планеты, районы H_{II} и некоторые галактические туманности принадлежат к числу возможных объектов изучения в этом диапазоне спектра.

Лунный день будет неблагоприятным для любых высокочувствительных наблюдений, поскольку солнце является доминирующим источником радиоизлучения в небе, особенно на коротких волнах. Основным объектом изучения в этот период будет само солнце и солнечная корона.

Introduction

The observing conditions under terrestrial, orbital and lunar environments differ considerably throughout the radio spectrum. A terrestrial radio astronomy observing station is limited at long waves by the presence of the ionosphere and atmospheric noise, as well as by man-made interference. Considerable absorption, refraction and faraday rotation is often encountered. The meter and decimeter range is relatively free from the above mentioned drawbacks with the exception of man-made interference. The centimeter and shorter waves are again considerably influenced by variable absorption and thermal radiation of water vapour and oxygen. This constitutes a serious limitation to the performance of sensitive radio telescopes. Further there is atmospheric refraction and man-made interference.

Earth orbiting radio telescope may considerably reduce the effects of the ionosphere and the atmosphere, yet, even then, ionospheric and exospheric noise is an important factor at long waves. Man-made interference will still be present almost throughout the whole radio spectrum.

The lunar environment, devoid of an atmosphere, compares favourably with a terrestrial or an Earth-orbiting radio astronomy observatory. The main lunar limitations are likely to be met at very long waves. There the lunar ionosphere, if present, and the solar wind or possibly the Earth tail and the interstellar ionized matter may be of importance. The opposite side of the Moon would not suffer from any terrestrial interference either natural or man-made. The main limitations of such a lunar observing site will most likely be due the to Sun, if interstellar ionization is relatively small. Thus the opposite side of the Moon during the lunar night should offer us one of the best observing sites that is to be found near the planets of the inner solar system.

Instrumental Problems of a Lunar Radio Astronomy Observatory

The wide radio spectrum available at the Moon poses great instrumental problems for lunar radio astronomy. To obtain resolution at kilometer wave-range dimensions of instruments of the order of hundreds of kilometers are required. Of all techniques available today, only the interferometer employing aperture synthesis seems to be able to cope with such difficult requirements. The obvious requirement of simplicity and low weight restricts the aerials to very simple constructions.

In order to reduce transportation problem for the kilometer wave aerials only one aerial could be based on the Moon. Two more would orbit the Moon on two mutually perpendicular orbits, and thus make the aperture synthesis at kilometer or longer waves possible. The aerials used should meet the wide relative band-width requirement and travelling-wave type aerials would be a reasonable choice with two receivers serving also as matched terminations to improve reliability and energy efficiency.

Laser or radio links would be used for information transfer from the aerials to the lunar station. Laser systems may be preferred in order to obtain high accuracy on directional and distance coordinates (laser radar and information channel). The information thus obtained would feed a computer performing both data logging and reduction.

The orbiting aerials could also possibly be used for hectometer waves, however fast changes of interference fringes may be the main limiting factor here. So possibly part of the hectometer spectrum as well as the decameter waves could best be served by a lunar-based aperture-synthesis system with moving aerials mounted on trolleys suspended on lunar mountains.

In this wave-length range log-periodic or rhombic aerials would be the most useful types. Beginning from meter waves downward, parabolic or spherical dishes would become advantageous. New constructional designs benefiting from the low lunar gravity and lack of winds should be attempted with special emphasis on easy assembly and transportation. It is conceivable that some very light and folding structures using plastic or metal three-dimensional cobweb-like frame structures inflated and stretched by polymerizing plastic foam could do the job.

The necessary reflecting surface could be previously deposited metal foil on plastic or a vacuum sprayed metal film coat applied at the lunar site. Metal foil coverage or vacuum metal spraying of the dish surroundings will reduce the lunar thermal radiation, and help to keep the aerial temperature low. For the same purpose the side-lobe level should be as low as possible. It may be worthwihle to use natural craters of proper shape as reflectors from decimeter up to hectometer waves though the feed positioning may pose some difficult problems. The centimeter and shorter waves down to perhaps the infrared range would be the other promising part of the spectrum. This part of the spectrum would most likely best be served by reflecting-type instruments. Consideration should be given to the required accuracy of the surface. If accurate dismountable dishes could be transported safely to the Moon a wide spectral range could be served by a small number of dishes. Accurate large dishes would also be advantageous at decimeter waves giving simultaneously high sensitivity, directivity and beam efficiency. It seems that the technique of aperture synthesis which achieves, extremely simply, directivity and sensivity at low cost on the Earth will also be the main lunar technique from kilometer down to perhaps centimeter waves.

The receiving equipment will most likely all be of solid-state type with proper circuit redundancy to increase reliability. The receiving and calibrating systems should be as fully automatic as possible with digital information output and control. For the higher frequencies, beginning with meter waves, cooled para-metric amplifiers will be the most useful, sensitive, simple and tunable amplifiers down to centimeter waves. They will possibly outperform masers at that time. Masers are likely to be important at cm and mm waves, though for some very wideband observations the liquid helium cooled germanium detectors may be of importance down to the infrared waves, Irrespective of the new techniques that may be developed before LIL becomes operational, it is obvious that all techniques will be photon noise limited, beginning from the decimeter range downwards.

Some Possible Types of Research at a Lunar Radio Astronomy Observatory

Lunar Nighttime

Very long wave observations will most likely be limited by the following: the absorption of radio waves by the lunar ionosphere, solar plasma, Earth tail or interstellar ionized matter. Further research is necessary to establish the relative role of these factors, and hence the scope of future radio astronomy investigations. If the dominating effect is due to the lunar ionosphere, lunar orbiting aerials will be the only way to go beyond the low frequency ionospheric limitations. If the Sun is the most important source of ionized gas, then periods of sunspot minima may be the only time when very long wave observations can be made or observing sites near remote planets might be required. Most serious would be interstellar ionized matter, since going beyond it would be outside the reach of space travel for a very long time to come.

Supposing that wavelengths up to about a kilometer will be available for observations on the Moon, then investigations may cover the following subject: Galactic and extragalatic spectral observations of low angular resolution to establish the frequency of maximum flux density and the shape of the low frequency part of the radio spectrum. This is important for theory and for the determination of the relativistic electron energy distribution.

It should be noted that extragalactric backgroung radiation may become dominant below the frequency of the maximum flux density of the galaxy. The increasing recession velocities of the progressively more distant radio galaxies will cause their frequencies of maximal flux densities to move to longer waves as their distance increases. The integrated extragalactic sky background should have a low frequency spectrum different from that of the galaxy. It may be possible to compare the observed and theoretical spectra for different cosmological models, and thus find the distribution of matter with increasing "redshift". This probably would be one of the simplest observations that can be made from the other side of the Moon or from a Moon orbiting simple aerial.

If we increase the resolution by means of aperture synthesis at hectometer and kilometer waves, we may be in the position to study the spectra of different parts of our galaxy and a large number of radio sources. It is important to study the spectra of a large number of extragalactic radio sources at very long waves. This cannot be done from the Earth, and it is very unlikely that this can be done from Earth satellites, due to the presence of ionospheric and exospheric noise and man-made interference.

By measurement of the spectra of numerous radio galaxies at the very low frequencies it should be possible to observe and measure their "redshift". This would give us a method of measuring their distance by radio observations alone. This is very important because the majority of radio sources are most likely to be beyond the limits of optical identification. It is essential that such spectral investigations do not suffer from any man-made interference on the Moon or in its vicinity. Special care should be taken to safeguard this vital low frequency part of the radio spectrum from harmful interference from any equipment used by men on the Moon or on lunar orbiting vehicles.

Additional information on galactic and radio source spectra in the hectometer and decameter range will be obtained by other Moon-based aperture synthesis aerials. Long wave observations during the lunar night will also be useful for solar-system studies (interplanetary weather), and especially for Jupiter spectral observations (Jupiter may be a source of serious interference for other observations).

Aperture synthesis techniques using steerable and movable dishes working in the meter down to centimeter wave ranges, which lie in the most transparent part of the terrestrial radio window, can also be successfully used on the Moon. A lunar observing site should have higher inherent sensitivities because of the increased available bandwidth and lower lunar thermal radiation. The absence of atmospheric and ionospheric irregularities and diffraction should also enable one to achieve higher angular resolution. The lunar environment will thus be favourable for 21 cm hydrogen line and OH line studies. Such refined observations as the Zeeman splitting of the 21 cm line, determined by the intensity of the interstellar magnetic field, is still another example. The lunar observing site should greatly help us in the achievement of the ultimate sensitivities possible in the decimeter and centimeter range.

Special care should be taken in the design of aerials to reduce the effects of side lobes and instrumental polarization and resulting changes of zero level during aerial scanning. Utmost sensitivity will possibly be achieved with stationary aerials scanned by the rotation of the Moon, using also multiple feeds in the focus.

The centimeter and shorter waves will be free from variable absorption caused on the Earth by atmospheric water vapour and oxygen, as well as by rain, clouds and snow. The moderate dish size needed for high resolution at these wavelengths may cause such dishes to be the first to be carried to the Moon. The main problem here is the achievement of sensitivity and not resolution. The best amplifiers will soon be photon noise limited (a fundamental limitation). The only way to increase sensitivity without an increase of resolution is to make observations with a number of identical dishes limited by photon noise, and to correlate the outputs or to repeat observations, if time is not important. It is difficult to see what could be studied in this frequency range. The stronger radio sources, the galactic center, the planets, H_{II} regions and some galactic nebulae are among the possible targets.

Lunar Daytime

The lunar day will considerably limit all high sensitivity observations. The quiet Sun will mostly influence the short wave ranges because its relatively high brightness temperature will be visible through the side lobes of the aerial. Thus the main advantage of high sensitivity is lost during lunar daytime.

The main object for studies at this time will be the Sun itself and the solar corona. Very low frequency solar spectral observations, as well as spectral and positional observations of solar radio bursts below the frequencies of terrestrial ionospheric opacity, can then be obtained. Other possible observations may be made of radio source occultation for studies of the outer solar corona and of long radio-wave propagation in the corona. Interplanetary scintillations and their influence on the maximum available resolution at long wave range during lunar daytime may also be investigated.

Solar burst propagation, its behaviour in interplanetary space and its influence on "interplanetary weather" are problems of tremendous importance to future space flights.

Solar radar investigations at low frequencies would also offer some interesting information about the behaviour of the solar corona and the solar wind.

Further topics for study include the lowest usable frequency on the Moon during daytime and its depence on the angle of the Sun in respect to the lunar horizon.

References

A. Hewish, P. F. Scott and D. Wills, Interplanetary Scintillation of Small Diameter Radio Sources. Nature 203, 1214 (1964).

F. J. Low, Performance of Thermal Detection Radiometers at 1.2 mm. Proc. I.E.E.E. 53, 516 (1965).

F. J. Malina, Report of the Lunar International Laboratory Discussion Panel, Warsaw. Astronaut. Acta 11, 123 (1965).

M. Ryle and A. Hewish, The Synthesis of Large Radio Telescopes. Monthly Notices Roy. Astronom. Soc. 120, 220 (1960).

Orhaug Torleiv, The Effect of Atmospheric Radiation in the Microwave Region. Publications of the National Radio Astronomy Observatory 1, No. 14, Green Bank, W. Va., October, 1962.

I. I. K. Pauliny Toth, J. R. Shakeshaft and R. Wielebinski, The Use of a Paraboloidal Reflector of a Small Focal Ratio as a Low-Noise Antenna System. Proc. I.R.E. 50, No. 12 (1962).

Eclipse Observation from the Moon and Cislunar Space[1]

By

G. F. Schilling[2] and **R. C. Moore**[2]

(With 4 Figures)

Abstract — Résumé — Резюме

Eclipse Observation from the Moon and Cislunar Space. The observation of lunar eclipses from the Earth's surface is quite frequent. The observation of solar eclipses, on the other hand, requires a special expedition to the specific geographical area where one is visible; such expeditions frequently are foiled by weather. Observed on Earth, the duration of totality of a solar eclipse is very short. Even on board a highspeed jet aircraft chasing the fleeting shadow above the clouds, there are only a few minutes in which to perform the most important experiments.

This study concentrates on the unique opportunities available for performing valuable scientific experiments on the Moon during the occurrence of eclipses of the Sun and Earth. Complementary observations that can be obtained concurrently from the Earth and from Earth-orbiting spacecraft are also examined. Two time phases are distinguished since progress in this transplanted field of a very old science depends directly on the growth of the space program. In the immediate future, scientific results will be limited and will pertain primarily to solar physics and geophysics. Nevertheless, it can be shown that even very simple observations obtained during the initial lunar landings will reveal astonishingly valuable scientific information.

For each of the two phases mentioned above, the study discusses the scientific experiments and observations that will probably be most fruitfully performed on the Moon, on Earth, and aboard spacecraft. The range of instrumentation and equipment extends from simple stop watches, cameras, and geigercounters to sophisticated structures such as radio telescopes on the Moon. Throughout the paper, it is shown that the value to be derived from the observations will be greatly enhanced by simultaneous observations from different locations.

The substantive results are essentially three-fold. First, the study illustrates how the obtaining of eclipse observations from extraterrestrial locations creates startingly unique opportunities to conduct a variety of scientific experiments not possible from Earth in a number of diverse disciplines. Secondly, it shows the importance of conducting coordinated, simultaneous observations both with regard to location on the Moon, on the Earth, and in space, and with regard to different disciplines. Finally, it emphasizes the value of international cooperation to such an eclipse program. This cooperation should encompass amateur as well as professional astronomical observations of eclipses from all parts of the earth, at the same time that space scientists perform their observations on the Moon and aboard spacecraft.

Observation des eclipses à partir de la Lune et de l'espace cislunaire. L'observation d'éclipses de lune est assez fréquente sur terre. L'observation des éclipses de soleil, d'autre part, exige des expéditions spéciales vers les zones géographiques d'où elles

---

[1] Any views expressed in this paper are those of the authors.
[2] The Rand Corporation, Santa Monica, California, U.S.A.

sont visibles; de telles expéditions échouent fréquemment à cause du mauvais temps. Observée de la terre, la durée totale d'une éclipse de soleil est très courte. Même à bord d'un avion à réaction rapide, donnant la chasse à l'ombre fuyante au dessus des nuages, on ne dispose que de quelques minutes pour exécuter les expériences les plus importantes.

Cette étude concentre son attention sur les possibilités particulières dont on dispose sur la lune pour faire des expériences scientifiques précieuses au cours des éclipses de soleil et de terre. On envisage aussi les observations complémentaires qui peuvent être faites concurremment de la terre et des engins gravitant autour de la terre. On distingue deux phases puisque les progrès dans ce domaine, dérivé d'une science très ancienne, dépend directement du développement du programme spatial. Dans le proche avenir, les résultats scientifiques seront limités et se rapporteront d'abord à la physique du soleil et de la terre. On peut montrer néanmoins que les observations même très simples obtenues lors du premier débarquement sur la lune, apporteront la révélation de renseignements scientifiques étonnamment précieux.

Cette étude discute, pour chacune des phases citées plus haut, les expériences scientifiques et les observations qui seraient faites avec le plus de profit, sur la terre, sur la lune et à bord d'un engin spatial. L'instrumentation et l'équipment vont du simple chronomètre, des appareils photographiques et des compteurs de Geiger à des montages élaborés tels que des radiotéléscopes sur la lune. Tout au long de cet article on montre que le profit tiré de ces observations sera fortement accru par des observations simultanées faites en différents lieux.

Les résultats substantiels sont essentiellement de trois sortes. Cette étude illustre d'abord comment le fait d'observer les éclipses hors de la terre fait naître dès le départ des possibilités exceptionnelles d'expériences scientifiques impossibles sur terre et ceci dans des disciplines diverses. Deuxièmement, elle montre l'importance de faire des observations simultanées, coordonnées à la fois par rapport aux lieux, sur la lune, sur la terre et dans l'espace, et aussi par rapport aux différentes disciplines. Enfin, elle insiste sur l'importance d'une coopération dans l'interprétation des résultats d'un tel programme. Cette coopération devrait comprendre l'observation astronomique de l'éclipse aussi bien par des amateurs que par des professionnels, sur tous les points du globe, pendant que les scientifiques de l'espace feraient leurs observations sur la lune et à bord d'un véhicule spatial.

Наблюдение затмений с Луны и из окололунного пространства. Наблюдения лунных затмений с поверхности Земли выполняются довольно часто. Наблюдение же солнечных затмений требует организации специальных экспедиций в определенные географические районы, где такое затмение можно наблюдать. Зачастую погода препятствует работе таких экспедиций. При наблюдении с Земли продолжительность солнечного затмения очень невелика. Даже с борта реактивного самолета, следующего над облаками за движущейся тенью Луны наблюдатель может выполнять важные научные эксперименты лишь в течение считанных минут.

Исследование касается исключительной возможности проведения ценных научных экспериментов на Луне, во время затмений Солнца и Земли. Рассматривается также вопрос о дополнительных наблюдениях с Земли и космических кораблей-спутников Земли. Можно усмотреть два периода этой работы, поскольку успехи перенесенной в новые условия столь старой науки, как астрономия непосредственно зависят от развития программ исследования космоса. В непосредственном будущем научные результаты будут ограниченными и будут касаться в первую очередь таких областей науки, как физика Солнца и геофизика. Тем не менее, можно указать, что даже самые простые наблюдения, выполненные при первых посадках на Луну, дадут чрезвычайно ценную научную информацию.

В исследовании разбираются, применительно к двум вышеупомянутым периодам, те научные эксперименты и наблюдения, которые, вероятно, дадут наилучший результат при выполнении с Луны, Земли и космического корабля. Потребуются разнообразные приборы и оборудование от простых часов, фотокамер и счетчиков Гейгера до таких сложных конструкций, как установленные на Луне радиотелескопы. Красной нитью через исследование проводится мысль о том, что ценность полученных наблюдений будет значительно выше, если одновременно будут проводиться наблюдения из нескольких разных точек.

Из исследования вытекают следующие три основных заключения. Во-первых, исследование показывает, каким образом наблюдение с внеземных пунктов открывает новую и единственную в своем роде возможность проведения разнообразных научных экспериментов по целому ряду дисциплин, выполнение которых на Земле невозможно. Во-вторых, оно показывает значение согласованного и одновременного проведения наблюдений как с разных точек — Луны, Земли, космического корабля, — так и в интересах разных дисциплин. Наконец, в нем подчеркивается значение международного сотрудничества в проведении такого рода программы изучения затмений. Это сотрудничество должно направлять внимание астрономов как любителей, так и профессионалов всех стран, на наблюдение затмений с Земли, в то же самое время, когда ученые-космонавты выполняют свои наблюдения с Луны или с борта космического корабля.

Introduction

Eclipses of the Moon and the Sun have been observed and studied since ancient times. In fact, it was not far from here that THALES OF MILETUS accurately predicted the solar eclipse of May 28, 585 B.C. Yet, some 2,500 years later, eclipse theory is still complex and wholly inadequate to explain all observed phenomena and effects (BARBIER [1], LINK [2]).

Observations of lunar and solar eclipses are enormously valuable to a number of scientific disciplines. But our observations have been severely limited to date by the geometry and motions imposed by the solar system. In the near future, man will be established in laboratories orbiting the Earth at great altitudes, in spacecraft passing to and from the Moon, and on the Moon itself. From these vantage points, he will be able to take advantage of unique opportunities to perform valuable scientific eclipse experiments heretofore impossible.

In a series of recent studies, R. C. MOORE and I have investigated some of these possibilities. We have analysed how a well-coordinated program of simultaneous eclipse observations from the Earth and those extra-terrestrial laboratories will enhance our knowledge of the Sun, the Moon, and the Earth itself [3, 4]. For example, using available data on the shape of the Earth's shadow during lunar eclipses, it is possible to infer characteristics of the Earth's upper atmposhere [5] in a region still rather inaccessible to other methods of investigation (KONDRATIEV [6]). Just recently, there was a calculation of the brightness of the Earth's halo as it would be seen from various distances within the terrestrial shadow-cone [7] — a first step towards a quantitative solution of how the Earth as a planet appears to a distant observer in space.

At this presentation, we wish to discuss two specific results of our investigations:

(a) The basic conditions that permit the advantageous conduct of eclipse observations by observers on the Moon and in cislunar space, and the uniqueness of the possible scientific experiments, and

(b) Quantitative values of the expected brightness of the terrestrial halo surrounding the Earth whenever an observer, regardless of his location, experiences a central total eclipse of the Sun with the Earth as the eclipsing or occulting body.

Basic Eclipse Geometry

Astronomical eclipses occur when the light from one celestial body is prevented by an intervening second body from reaching the observer. Now the observer may be located on the second or a third celestial body or in a spacecraft. It is this location that determines the nature as well as the duration of an eclipse. The basic geometry for the Earth-Moon system is schematically illustrated in Fig. 1.

On Earth, we are all familiar with both the lunar eclipse, when the Moon passes through the Earth's shadow, and the solar eclipse, when our view of the Sun is obscured by the intervening Moon. Note that someone on the Moon would be observing a *solar eclipse* when we on Earth would be observing a lunar eclipse. Similarly, a *terrestrial eclipse*—really a lunar-shadow transit—is seen from the Moon when a solar eclipse is observed on earth.

Let us investigate further the significance of the basic celestial geometry by examining the phenomenon of the solar eclipse. On Earth, eclipses of the Sun

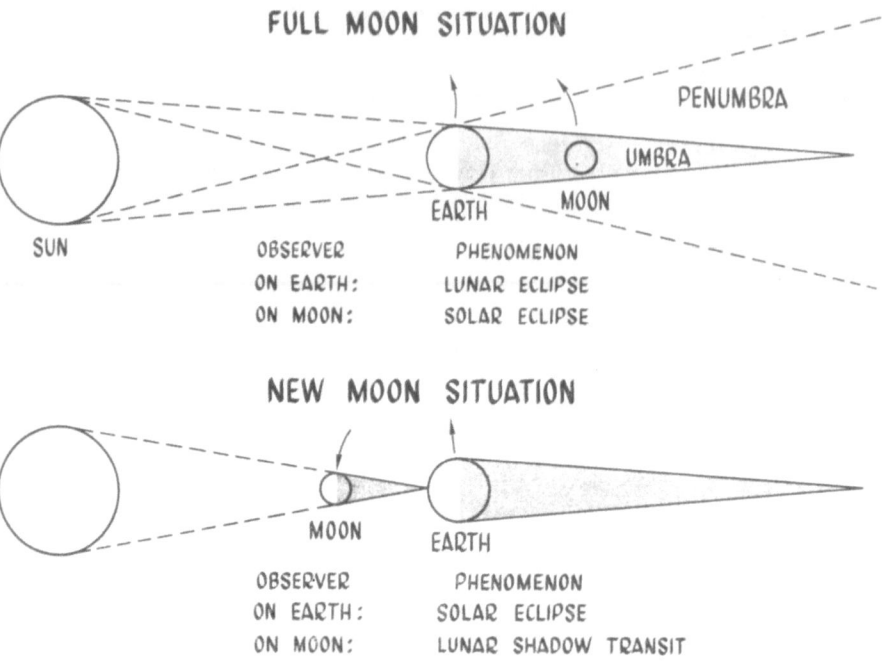

Fig. 1. Schematic illustration of the basic eclipse geometry involving the Earth and Moon (not to scale)

are more frequent than lunar eclipses, but each one is observable only from a very limited area of the Earth's surface. The preponderance of oceans, difficulty of access to many land regions, and vagaries of weather conditions further limit the chances of obtaining useful data. When all these limitations have been overcome, the duration of totality at a station near the equator, for instance, may be at most $7^m 40^s$.

In strong contrast, a solar eclipse observed on the Moon will be quite an extensive phenomenon, both spatially and temporally. The duration of totality will be as much as 2-1/2 hours, and the eclipse will be visible for a relatively long time from all points of the lunar hemisphere for which the Earth is above the horizon. Furthermore, the usual atmospheric interference with observations will be missing, since the Moon has no appreciable atmosphere (Zwicky [8]). On the other hand, we can expect the Earth to be surrounded by a narrow, bright fringe caused by multiple scattering of sunlight in the earth's atmosphere.

Eclipses follow a precise time schedule. Table I is a list of the eclipses that will occur during the next few years, after the classical canon by Oppolzer [9].

Table I. Eclipse Schedule

Solar Eclipse on Moon (Lunar Eclipse on Earth)		Terrestrial Eclipse on Moon (Solar Eclipse on Earth)
1966	none	20 May * Africa—Asia 12 November * South Amerika
1967	24 April * 18 October *	9 May 2 November * South-Atlantic
1968	13 April * 6 October *	28 March 22 September * Asia
1969	None	18 March 11 September
1970	21 February 17 August	7 March * Central America 31 August
1971	10 February * 6 August *	25 February 22 July 20 August
1972	30 January * 26 July	16 January 10 July * North America
1973	10 December	4 January 30 June * Africa 24 December
1974	4 June 29 November *	20 June * Australia 13 December
1975	25 May * 18 November *	11 May 3 November

* Total eclipse. Principal areas for visibility on Earth are indicated for total solar eclipses.

Eclipse Observations from the Moon

The general scientific aims and objectives of eclipse research are well known, and a variety of techniques have been developed for making measurements. Considering the unique advantages, no one should doubt that eclipse observations from the Moon will eventually contribute to considerable advances in astronomy, astrophysics, and geophysics, as well as many areas of astronautics and even relativity theory. There are, however, two additional aspects to be considered.

First, human observers can conduct valuable scientific experiments in this field with very simple instruments. Thus, even the earliest manned landings on the Moon could provide the beginning of an extensive program of eclipse studies. Operational problems are minor, except for the necessity to time missions around the one to four favorable periods in each year.

Secondly, the value of scientific data can be greatly enhanced by planning a program of simultaneous observations from the Moon, from Earth, and from

Table II. *Initial Experiments During the Occurrence of a Solar Eclipse on the Moon*

Observing Site	Observational Program
On the Moon	Color photographs of Sun Time-lapse movies of Solar corona Timing of eclipse progress Filter photography of Sun and solar corona Timing of shadow passage at observing site Soil temperature at observing site Cosmic-ray intensity at observing site Color photography of terrain of observing site (shadow intensity)
On Earth	Standard professional observations for occurrence of lunar eclipse (astronomy, meteorology, lunar physics) Amateur astronomer participation (photography, eclipse timing, shadow brightness, crater timings) Public participation (local weather information, photography, timings)
Aboard Spacecraft (in Earth Orbit)	Photography of Moon Photometry of albedo Weather formations along terminator

Table III.
Initial Experiments During the Occurrence of a Terrestrial Eclipse on the Moon

Observing Site	Observational Program
On the Moon	Photography of Earth Time-lapse movies of shadow progression Timing of shadow transit over Earth Albedo observations Radio reception studies
On Earth	Standard professional observations for occurrence of solar eclipse (astronomy, solar physics, ionospheric physics, radio astronomy) Amateur astronomer participation (photography, eclipse timing) Public participation (solar photography, timing, terrain photography, radio reception, air temperature measurements)
Aboard Spacecraft (in Earth Orbit)	Photography of Sun and solar corona Radio transmission and reception studies Photography of shadow transit on Earth Cosmic-ray experiments

Earth-orbiting spacecraft. Tables II and III list suggested initial observations during a solar and a terrestrial eclipse, respectively. These lists, though not exhaustive, do indicate the kind of simple experiment suitable for early, exploratory flights to the Moon. All of the studies mentioned require only uncomplicated photographic equipment with various filters, thermometers, a simple telescope, a stop watch, a geiger counter, and similar instruments.

During later phases, much more sophisticated equipment would be desirable. Eventually, one can foresee demands ranging from large telescopes and Schmidt cameras to radio telescopes. Once a lunar scientific laboratory becomes feasible (MALINA [10]), a program of eclipse observations will certainly expand to make use of such opportunities as the occultation of stars by planets, or the study of eclipsing binary stars.

Eclipse Observations from Spacecraft

So far, we have been discussing only eclipses as they occur in nature determined by the relative motions of Earth, Moon, and Sun.

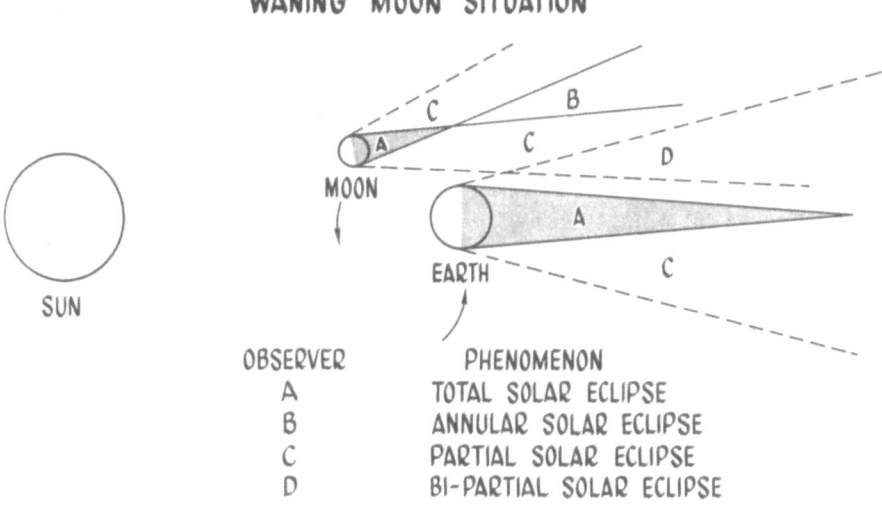

Fig. 2. Schematic illustration of the basic eclipse geometry involving the Earth, Moon and spacecraft (not to scale)

Fig. 2 is indicative of the further opportunities that are available for eclipse research when one includes manned spacecraft, either in orbit around the Earth or in cislunar space. Tables IV and V list the principal eclipses, transits and occultations involving the Sun, Earth, Moon and an orbiting spacecraft. Each time the spacecraft observer passes through the umbra or penumbra of either the Earth or the Moon, a number of phenomena become observable. The frequency of eclipses and occultations increases extraordinarily once they can be arranged artificially through choice of proper orbits. Mathematically, this is the result of adding a fourth body—a satellite or spacecraft—to the Earth-Moon-Sun system, thereby allowing the location of observers on any of the three smaller bodies involved. Scientifically, the research opportunities so created are challenging.

Table IV. *Selected Principal Eclipses, Transits and Occultations Involving the Sun, Earth, Moon and a Spacecraft Orbiting in the Earth's Shadow*

Location of Artificial Satellite (in orbit)	Observing Site	Object Observed	Phenomenon
Between Earth and Moon during lunar eclipse	Earth	Moon Satellite	Lunar eclipse Satellite eclipse Satellite transit of Moon
	Moon	Earth Satellite	Solar Eclipse Satellite transit of Earth Satellite transit of Sun
	Satellite	Moon Earth	Lunar eclipse Solar eclipse or terrestrial occulation of Sun
Between Earth and Moon's distance during solar eclipse	Earth	Moon Satellite	Solar eclipse Satellite eclipse
	Moon	Earth Satellite	Lunar shadow transit of Earth Terrestrial occultation of satellite
	Satellite	Moon Earth	Terrestrial occulation of Moon Solar eclipse or terrestrial occultation of Sun
Beyond Moon during lunar eclipse	Earth	Moon Satellite	Lunar eclipse Satellite eclipse Lunar occultation of satellite
	Moon	Earth Satellite	Solar eclipse Satellite eclipse
	Satellite	Moon Earth	Lunar transit of earth Solar eclipse
Beyond Moon's distance during solar eclipse	Earth	Moon Satellite	Solar eclipse Satellite eclipse
	Moon	Earth Satellite	Lunar shadow transit of Earth Terrestrial cocultation of satellite
	Satellite	Moon Earth	Terrestrial occultation of Moon Solar eclipse

The Earth's Halo

Present eclipse theory is quite inadequate for quantitatively explaining a number of phenomena observable during an eclipse. A major deficiency is our present inability, even with advanced computing techniques, to solve rigorously problems of radiation transfer in a real atmosphere — specifically the propagation of sunlight through the Earth's non-homogeneous atmosphere, affected by preferential absorption, multiple scattering, and differential refraction.

This influence of the Earth's atmosphere is present whenever the Earth is the eclipsing or occulting body, that is, during all solar eclipses observed from the Moon or from a spacecraft. The theoretical inadequacies represent a severe problem, therefore, in developing an eclipse research program that will realistically plan the data analysis.

We have obtained a partial solution based on a semi-empirical method that combines refraction and multiple-scattering theory with observed measurements

Table V. *Selected Principal Eclipses and Transits Involving the Sun, Earth, Moon and a Spacecraft in the Moon's Shadow*

Location of Spacecraft (in orbit)	Observing Site	Object Observed	Phenomenon
Between Earth and Moon during solar eclipse	Earth	Moon Satellite	Solar eclipse Satellite eclipse in Moon's shadow Satellite transit of Moon
	Moon	Earth Satellite	Lunar shadow transit of Earth Satellite eclipse in Moon's shadow Satellite transit of Earth
	Satellite	Moon Earth	Solar eclipse *or* lunar occultation of Sun Lunar shadow transit of Earth
Between Earth and Moon near New Moon	Earth	Satellite	Satellite eclipse in Moon's shadow Possible transits of Moon and Sun or both
	Moon	Satellite	Satellite eclipse in Moon's shadow Possible satellite transit of Earth
	Satellite	Moon	Solar eclipse *or* lunar occultation

of the brightness of the twilight sky. The method treats the atmosphere itself as a luminous source [11, 12], and is described elsewhere [4, 7]. Even with modern computing machines, the calculations are tedious and time-consuming. To date, we have successfully obtained quantitative data on the brightness of the Earth's halo as seen from altitudes out to the refracted vertex of the terrestrial shadow-cone.

Fig. 3 depicts this altitude regime that ranges from the surface of the Earth to a radial distance of 255,000 km. The inset illustrates the general appearance of the Earth, to be visualized as hiding the Sun from view during a central total eclipse observed from a spacecraft. The dark Earth is surrounded by a luminous fringe, caused by refraction and multiple scattering of sunlight in the atmosphere, attenuated both by scattering and by absorption. The relative angular extent of the contributing atmosphere is about 1/60 of the radius of the Earth. The variation of the absolute angular values A with radial distance is given in the diagram, together with the angular extent B of that portion of the scattering atmosphere that contributes more than 70% towards the luminous appearance of the halo.

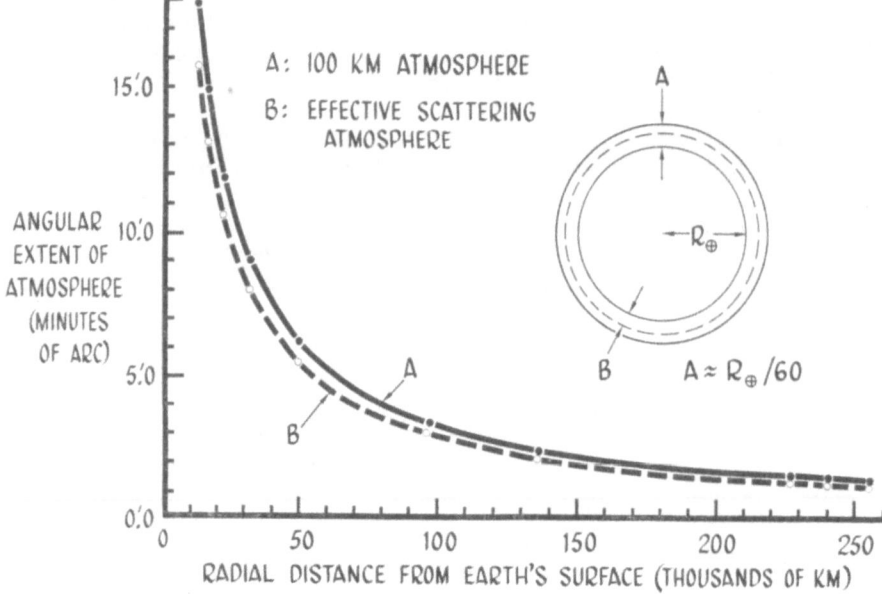

Fig. 3. Variation of apparent angular extent of Earth's atmosphere (A) and luminous halo (B) with radial distance

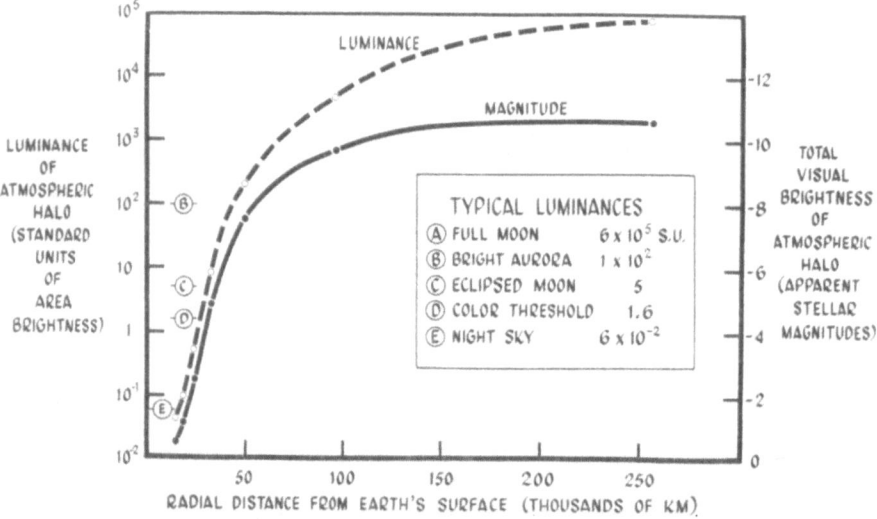

Fig. 4. Variation of visual surface brightness (luminance) and total brightness (magnitude) of Earth's halo with radial distance

Fig. 4 shows absolute values of the visual brightness of this atmospheric fringe region. The total brightness of the whole halo ring, labelled *Magnitude*, for average meteorological conditions in the area of the terrestrial terminator, is given in conventional astronomical units of apparent stellar magnitude. The major contribution in terms of *Luminance* or surface brightness comes from the inner portion of this ring (cf. curve *B* in Fig. 3). It extends from the apparent surface of the Earth to an angular distance corresponding to about 10° slant elevation relative to the tangential point of the limiting solar rays [7].

A word about the units used. Photometric as well as radiometric measurements make use of a confounding array of dimensions and units [13]. For our present problem, we were concerned mainly with area brightness. In Fig. 4, therefore, the luminance or visual surface brightness of the atmospheric inner fringe is given in adopted standard units of area brightness (s.u.), defined as the number of 0-magnitude stars per square degree of solid angle. This unit can perhaps be best understood in terms of personal experience (cf. Table VI below):

Table VI. *Typical Values of Surface Brightness*

Phenomenon	Approximate Standard Units of Area Brightness	
Night Sky	0.06	s.u.
Color Threshold of Human Eye	1.6	s.u.
Eclipsed Moon	5	s.u.
Bright Aurora	1×10^2	s.u.
Zenith Sky at 3° Solar Depression	3×10^3	s.u.
Full Moon	6×10^5	s.u.
Sun	2×10^{11}	s.u.

The basic relation of this standard unit to other units of visual surface brightness, commonly used in photometry, is given below:

1 s.u. $= 7.8595 \times 10^{-7}$ candles cm$^{-2} = 2{,}469.135$ milli-micro-Lamberts;

$$\log_{10} n = -0.4 \, m;$$

$$m = 2.52 - 2.5 \log_{10} [b \times 3{,}600^2]$$

$$= -14.052 - 2.5 \log_{10} [b \times 10^4 \times (\pi/180)^2]; \text{ and}$$

$$M = -2.5 \log_{10} [n \times S],$$

where *n* is the area brightness in standard units (s.u.), that is, the number of 0-magnitude stars per square degree of solid angle, and where *b* is the surface brightness in candles per square centimeter, *m* is the apparent visual stellar magnitude per square degree of solid angle, *M* is the apparent visual stellar magnitude of the object viewed, *S* is the solid angle of the object viewed in square degrees, and the apparent visual stellar magnitude of one lux is taken as -14.052 in accord with DE VAUCOULEURS' most recent determination [14].

Fig. 4 reveals an interesting phenomenon. The brightness of the Earth's halo is considerably greater than would perhaps have been expected *a priori*; this is a direct result of considering multiple scattering in our calculations.

The data contained in Fig. 4 permit quantitative predictions of the appearance of the Earth to an observer (or appropriate instrument) located within the terrestrial shadow-cone, and are applicable to spacecraft in satellite orbits around the Earth as well as in cislunar space. The data also allow computations of the amount of visible light scattered into the geometric shadow-cone. Note that near a 255,000-km altitude, the area brightness of the inner fringe has almost reached that of the full Moon.

Beyond this altitude, the refracted though strongly attenuated image of the

Sun's rim will begin to appear to the observer as a bright ring of fire around the earth. This phenomenon will be equivalent to a change from a central total eclipse of the Sun to a central annular eclipse. We are presently engaged in extending our calculations into this regime and hope to have quantitative results soon. These data will be directly useful in specifying detailed instrumental requirements for eclipse observations from the surface of the Moon. Our preliminary results seem to indicate that a solar eclipse will indeed be a colorful sight for lunar observers. We also hope to be able to clarify several poorly understood problems associated with lunar eclipses, such as the apparent relation between the brightness of the eclipsed Moon and solar activity (Danjon [15], de Vaucouleurs [16]), the shadow increase phenomenon (Link [17]), and the excess flattening of the Earth's shadow [5, 18].

Conclusions

We have attempted to show that a program of eclipse observations from extraterrestrial locations will be an exciting field of research with unique opportunities to conduct scientific experiments not possible from Earth. It was relatively easy to outline the kinds of instruments needed and to stress how important it will be to coordinate simultaneous observations, with regard both to different disciplines and to the different locations on the Moon, Earth, and spacecraft.

There is nothing new *per se* about a suggestion to observe eclipses from the Moon. In fact, basic eclipse theory uses a mathematical approach along these lines (Link [19]). But eclipse theory is complex and often permits only approximate solutions. Certain radiation-transfer problems especially will need quantitative solutions [20] before proper use can be made of the data to be obtained. Although we have shown a few examples of such quantitative results, many more problems must be solved.

Ultimately, the success of extraterrestrial eclipse studies will depend on international cooperation among scientists of many disciplines. But in the past, major contributions to our knowledge of eclipses have resulted from observations made by amateur astronomers. Lunar and solar eclipses, unlike many scientific phenomena, are observable and therefore understandable to the non-specialist. Few adults have not had the opportunity to observe a lunar eclipse. Many have had the rarer but more awesome experience of a solar eclipse, with the Moon slowly obscuring the Sun until the whole landscape is bathed in dim twilight, and the spectacular solar corona makes its appearance in the starry sky.

For some time to come, opportunities for eclipse studies from extraterrestrial locations will be few and expensive. Our analysis indicates that professional experiments could be usefully supplemented by simple amateur observations, if conducted on a worldwide scale. Such coordinated efforts could greatly enhance the potential value of the professional data. We believe, therefore, that plans for an eclipse program should include invitations to the general public to participate at the same time that human observers are performing their observations on the Moon and aboard spacecraft.

References

1. D. Barbier, Photometry of Lunar Eclipses, Ch. 7, in: Planets and Satellites, G. P. Kuiper and B. M. Middlehurst, eds. Chicago: University of Chicago Press, 1961.
2. F. Link, Eclipse Phenomena, in: Advances in Astronomy and Astrophysics, Vol. 2, Z. Kopal, ed. New York: Academic Press, 1963.

3. G. F. SCHILLING, A Suggested Program of Eclipse Observations from the Moon. RM-4249-NASA, The RAND Corporation (1964).

4. R. C. MOORE and G. F. SCHILLING, Eclipse Observations from Orbiting Spacecraft. RM-4557-PR, The RAND Corporation (1965).

5. G. F. SCHILLING, Latitudinal Variation of Mesopause Height Inferred from Eclipse Observations. J. Atmos. Sci. **22**, 110 (1965).

6. K. YA. KONDRATIEV and O. P. FILIPOVICH, Teplyy Rezhim Verknikh Sloyev Atmosfery. Leningrad: Gidrometeorologicheskoye Izdatel'stvo, 1960.

7. R. C. MOORE and G. F. SCHILLING, On the Apparent Brightness of the Earth's Halo. The RAND Corporation (in preparation, 1965).

8. F. ZWICKY, Astronomy and Physics of the Moon. Lunar International Laboratory Committee No. 11, International Academy of Astronautics, Paris (1964).

9. T. VON OPPOLZER, Canon of Eclipses, Imperial Academy of Sciences, Vienna, 1887, translated by O. GINGERICH. New York: Dover Publications, Inc., 1962.

10. F. J. MALINA, Report of the Lunar International Laboratory Discussion Panel, Warsaw. Astronaut. Acta **11**, 123 (1965).

11. M. J. KOOMEN, C. LOCK, D. M. PACKER, R. SCOLNICK, R. TOUSEY and E. O. HULBURT, Measurements of the Brightness of the Twilight Sky. J. Opt. Soc. Amer. **42**, 353 (1952).

12. R. A. RICHARDSON and E. O. HULBURT, Sky-brightness Measurements near Bocaiuva, Brazil. J. Geophys. Res. **54**, 215 (1949).

13. I. J. SPIRO, R. CLARK JONES and D. Q. WARK, Atmospheric Transmission: Concepts, Symbols, Units and Nomenclature. Infrared Physics **5**, 11 (1965).

14. G. DE VAUCOULEURS, Geometric and Photometric Parameters of the Terrestrial Planets. Icarus **3**, 187 (1964).

15. A. DANJON, Sur une Relation entre l'Eclairement de la Lune Eclipsée et l'Activité Solaire. Comptes Rendus **171**, 1127 (1920).

16. G. DE VAUCOULEURS, La Loi Normale de Luminosité des Eclipses de Lune de 1894 à 1943. Comptes Rendus **218**, 655 (1944).

17. F. LINK, L'Absorption Atmosphérique et quelques Phénomènes Connexes. Bull. Observ. Lyon **11**, 229 (1929).

18. J. ASHBROOK, Measuring the Earth's Shadow. Sky and Telescope **27**, 156 (1964).

19. F. LINK, Lunar Eclipses, in: Physics and Astronomy of the Moon, Z. KOPAL, ed. New York: Academic Press, 1962.

20. Z. SEKERA and W. VIEZEE, Distribution of the Intensity and Polarization of the Diffusely Reflected Light Over a Planetary Disk. R-389-PR, The RAND Corporation (1961).

On Solar Observations at an International Observatory on the Moon

By

V. A. Krat[1]

(With 2 Figures)

Abstract — Résumé — Резюме

On Solar Observations at an International Observatory on the Moon. The main problem of solar observations on the Moon should be the continuous observation of solar activity in the region of γ-emission, X-rays and distant ultraviolet emission. The observation of separate spectral lines in the ultraviolet part of the spectrum with sufficiently high resolving power would permit the study of the active processes on the Sun simultaneously in the chromosphere, the intermediate zone between the chromosphere and the corona and also in the inner and outer corona; we consider these observations to be the most important, though the most difficult as regards techniques.

Instruments for obtaining information on γ-emission and the X-rays (monochromators) have been described in detail in the literature.

A complete set of instruments for investigation of the inner corona, the intermediate zone between the chromosphere and the corona and also the chromosphere should consist of: a photoelectric spectrometer of high resolving power — a double-pass monochromator designed by V. N. KARPINSKY ("Izvestia GAO", no. 178) and a battery of 20 small-sized spectroheliographs of A. V. MERKULOV's construction ("Izvestia GAO", no. 162, 1958) or modified versions of them. They are to be set on two lines of the solar corona, ten lines of the intermediate zone and six chromospheric lines. At the L α line the setting is to be made on the center and wings of the line.

It is expedient to install two coronographs for observation of the outher and inner corona in order to photograph the corona in white light.

If necessary, an X-ray counter as well as a scanning monochromator of the H α emission can be used in order to give a signal of the appearance of chromospheric flares. It is possible to use small lightened radio telescopes for the dcm-wavelengths.

L'observation du Soleil dans un Laboratoire International sur la Lune. Le but principal de l'observation du soleil sur la lune doit être une observation continue de l'activité solaire dans les régions du rayonnement γ, des rayons X et de l'ultraviolet lointain. L'observation de raies spectrales isolées dans la partie ultraviolette du spectre, avec un pouvoir de résolution suffisamment grand nous paraît particulièrement importante quoique la plus difficile du point de vue technique. Ces observations permettraient de suivre les processus actifs sur le soleil, simultanément sur la chromosphère, entre la chromosphère et la couronne, dans les couronnes interne et externe du soleil.

Les instruments destinés à donner des renseignements sur le rayonnement γ et les monochromateurs à rayons X ont été décrits suffisamment en détail dans la littérature.

L'ensemble complexe d'appareils destinés à l'étude de la couronne interne, de la zone intermédiaire entre la chromosphère et la couronne et de la chromosphère doit

---

[1] Pulkovo Main Observatory, Leningrad, U.S.S.R.

consister en: un spectromètre photoélectrique de grand pouvoir de résolution, un spectromètre à double diffraction de V. N. Karpinsky (Izvestia GAO n° 178) et une batterie de 20 spectrohéliographes de petit format de A. V. Merkulov (Izvestia GAO n° 162, 1958) ou une forme modifiée qu'il faut mettre au point sur deux raies de la couronne solaire, 10 raies de la zone intermédiaire et 6 raies de la chromosphère. Pour la raie Lα le réglage sera fait sur le centre et sur les ailes de raie.

Il est aussi utile d'installer deux coronographes pour la couronne interne et pour la couronne externe, afin de photographier la couronne en lumière blanche.

En cas de nécessité on peut faire usage du compteur à rayons X ou du monochromateur explorant l'émission de la raie Hα pour signaler l'apparition d'explosions chromosphériques. Il est possible d'utiliser de petis radio télescopés allégés dans la gamme décamétrique.

О наблюдениях солнца на международной обсерваторий на Луне. Основной задачей наблюдений Солнца на Луне должно являться непрерывное слежение за солнечной активностью в области γ — излучения, рентгеновского излучения и далекого ультрафиолетового излучения. Особенно важными, хотя и наиболее трудными в техническом отношении представляются нам наблюдения отдельных спектральных линий в ультрафиолетовом участке спектра при достаточно высокой разрешающей силе, которые позволяют проследить активные процессы на Солнце одновременно в хромосфере, переходной зоне от хромосферы к короне, во внутренней и во внешней короне Солнца.

Приборы для получения информации о γ — излучении и рентгеновские монохроматоры достаточно подробно описаны в литературе.

Комплекс приборов для изучения внутренней короны, — переходной зоны от хромосферы к короне и хромосфере должен состоять из фотоэлектрического спектрометра высокой разрешающей силы — спектрометр двойной диффракции В. Н. Карпинского (Известия ГАО № 178) и батареи из 20 малогабаритных спектролегиографов А. В. Меркулова (Изв. ГАО № 162, 1958) или их модификации, устанавливаемых на две линии солнечной короны, 10 линий переходной зоны и 6 хромосферных линий. У линии Lα установка производится на центр и на крылья линии.

Целесообразно также установить два коронографа для внешней и для внутренней короны для фотографирования короны в белом свете.

Для сигнализации о появлении хромосферных вспышек в случае необходимости может употребляться как счетчик рентгеновского излучения, так и сканирующий монохроматор в из..учении Hα. Возможно использование небольших облегченных радиотелескопод для дециметрового диапазона.

The main task of solar observation outside the Earth's atmosphere is obtaining information on the spectra of various solar formations (spots, faculae, flares and others) in short wavelengths — in the region of ultraviolet and X-ray emission. The most interesting spectral lines of the solar corona, chromosphere and intermediate zone between the chromosphere and the corona are just in this part of the spectrum.

According to the theory of the chromosphere and the corona, which is being worked out at Pulkovo, these outer parts of the solar atmosphere are extremely non-uniform. They consist of separate filaments and streams of plasma with different temperature, density, conductivity and magnetic field strength. This occurs when the chromosphere as well as the corona is not in equilibrium state. Fast motion, called, conditionally, solar turbulence, results in a continuous interchange between the photosphere, chromosphere and corona. This motion, in its turn, is being formed and maintained by non-uniformities in the magnetic field, which is organized during the process of solar activity. However, the transition of plasma from the coronal state (with a temperature of the order of 1 million degrees Kelvin) to the chromospheric state (where the temperature varies from 5,000° K to several tens of thousands of degrees), and in the opposite direction, cannot be

observed by observatories on the Earth. This is because the lines of ions of the
intermediate state are located almost entirely in the far ultra-violet part of the
spectrum. Therefore a series of physical processes, which are important for com-
prehending the nature of solar activity, escaped the astronomers' notice.

The main centers of formation of these non-uniformities are in the active
regions on the Sun and, in the first instance, in the huge magnetic fields of sunspots.
The most striking phenomenon of solar activity are the flares; during this process
a great amount of magnetic energy is changed into heat and mechanical energy.
A great portion of energy also passes into the resulting corpuscular streams, which
are called at present by the unsuitable term "solar wind". The streams of relativistic cosmical particles usually contain the greatest energy.

For the investigation of this interesting phenomenon it is not enough to make
observations by means of sporadic launchings of rockets and from artificial
satellites of the Earth. It is necessary to organize a daily observing service of
solar activity, and it would be expedient to make it on the Moon in order to
observe the Sun from the Moon's cloudless sky.

For continuous observations of the Sun it would be best to have two observatories on the Moon: one on the side visible from the Earth and the other on the
reverse side of the Moon.

At a lunar observatory the following problems of solar investigations could
be solved:

1. Investigations of the spectrum of the active regions on the Sun in the
region of the far ultra-violet part of the spectrum from 1,600 Å to 500—600 Å;
and,

2. Investigations of the solar spectrum in the X-ray and γ-ray regions.

The second problem can be solved with the help of instruments being used at
present for analogous purposes, that is, new or modified instruments need not be
designed. Such instruments for the measurement of X- and γ-ray emissions are
used in artificial satellites of the Earth and in rocket probes. They are required to
supply us with the spectra of X- and γ-ray emissions at different wavelengths,
and also with a rough (with a resolution of the order of 1' or 30") image of the
Sun in the X-ray emission region. In the future, the modification of A. V. MER-
KULOV's monochromators [1] will permit us to obtain images of the Sun in the
X-ray region with a considerably larger resolution.

It is very important to design a complex of instruments for observations of the
far ultra-violet part of spectrum (the first task). For starting this work, 18 spectral
lines can be selected: 2 coronal, 6 chromospheric lines and 10 lines of intermediate
state. The list of these lines is given in Table I.

The lines in brackets will be observed together. At L_α three monochromatic
sections will be cut out. One section near the center and two at the wings of the
line. Together with data on the lines L_β, L_δ and L_ε it would be possible to obtain
sufficiently reliable effective heights (above the photosphere) of the emission of
LAYMAN's hydrogen lines. Thus the development of active processes would be
observed from the chromosphere to the corona. Observations of the lines of the
intermediate zone will permit an evaluation to be made of the changes of temperature of the plasma and also of its electronic density. The whole physical process
can be observed in detail.

The most suitable monochromators for investigations in space are A. V. MER-
KULOV's monochromators of different type. The principle of operating these monochromators is described in Ref. [1]. These instruments differ from all existing
instruments because they have a dispersion unit in the focus of the objective
giving the image, and also in the objective forming the image on the exit slit

(Fig. 1). Since a high dispersion is required when operating at optical wave-lengths, it is necessary to use the FABRY-PEROT etalon combined with a diffraction grating. For the short-wave part of the spectrum only a diffraction grating can be used after the monochromator has been appropriately modified.

The MERKULOV monochromators are compact, portable and reliable in operation because there are no moving parts. It is possible to produce a number of

Table I

Element	Wavelength λ	Element	λ	Element	λ
\multicolumn{2}{c}{Corona}		\multicolumn{2}{c}{Intermed. state}		\multicolumn{2}{c}{Chromosphere}	
Ne VIII	770.5	C II	$\begin{pmatrix}1334.5\\1335.7\end{pmatrix}$	H	$\begin{matrix}L_\alpha\\L_\beta\end{matrix}$
Mg X	609.7	C III	977.0		
		,,	1176.0		L_δ
		C IV	1548.2		L_ε
				O I	$\begin{pmatrix}1304.9\\1306.0\end{pmatrix}$
		N II	1085.0		
		N V	1238.8	He I	584.3
		O V	629.7		
		O VI	1031.9		
			1037.6		
		Si IV	1393.7		

them which when joined to small concave mirrors, whose diameters are of the order of 10 cm, will form a battery of instruments giving images of the Sun in all the lines of interest to us. It is advisable to make 20 monochromators. They would

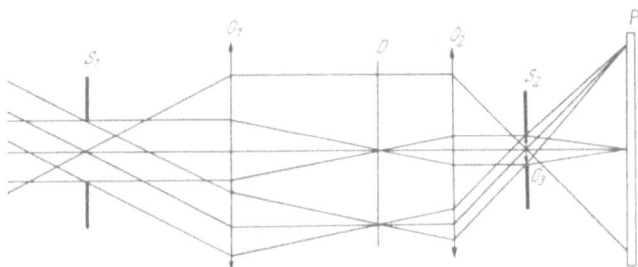

Fig. 1. MERKULOV monochromators
S_1 — entrance slit; O_1 — first objective; D — dispersion unit; O_2 — second objective; S_2 — second slit

require 20 film bobbins. A technical difficulty would possibly arise in connection with the rapid change of exposures and bobbins in vacuum. The development and reduction of the photographs should be done on the Moon.

To the battery of monochromators it would be expedient to add the V. N. KAR-PINSKY photoelectric spectrometer [2], which is the most precise of all present-day spectrometers. This is a double-pass monochromator which completely excludes scattered light from the spectrum records. In this instrument there is one moving part: the frame of the mirror ecker, reflecting the dispersed beam of light back to the grating (Fig. 2). Here, since we want to obtain a precise atlas of spectral lines, and to measure the intensities of faint spectral lines in sunspots and flares with great precision, it is necessary to secure a precise motion of the frame. In order to measure the whole spectrum from 560—600 Å to 1,600 Å, it would be necessary to change gratings, and to make records in two stages. It is evident that the KARPINSKY monochromator would cause considerable trouble to the observers in the Moon observatory, however, the obtaining of precise spectrophotometric solar data in the ultra-violet part of the spectrum would compensate all the trouble taken.

In addition to these main instruments, two coronographs need to be used for taking photographs of the inner and outer corona in white light. These coronographs are quite simple; they are prototypes of the LYOT coronograph for the

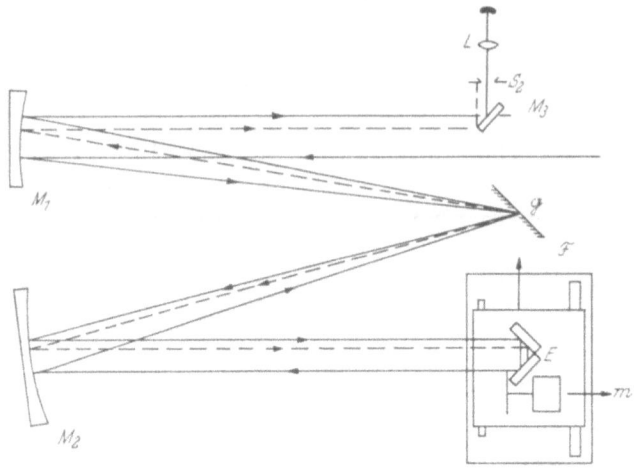

Fig. 2. Frame of the mirror ecker
PC — photomultiplier; S_2 — entrance slit; g — grating; E — ecker

inner corona and an externally screened coronograph for the outer corona. If the chromospheric flares should be dangerous for the observers, it would be necessary to use X-ray counters attached to the monochromator in the hydrogen line H_α, and also, in parallel with it, a simplified telescope for the dcm-wavelength range. These instruments would be connected to a signal system giving warning of danger.

At the beginning the solar equipment of a Moon observatory would consist only of these instruments. Later it would be desirable to mount on the Moon a large solar telescope with a coelestat mirror with a size diameter of 3—4 meters which would allow the use of its high resolution unhindered by an atmosphere. However, such an installation is still far from being solved technically, and at present, I feel that it is untimely to discuss other possibilities of the development of astronomy on the Moon.

References

1. A. V. MERKULOV, Pulkovo Observatory Publication, No. 162 (1958).
2. V. N. KARPINSKY, Bull. Solar Data, No. 1 (1961).

Exploration of the Interplanetary Medium from Lunar and Cislunar Space

By

R.-H. Giese[1] and R. Lüst[1]

(With 3 Figures)

Abstract — Résumé — Резюме

Exploration of the Interplanetary Medium from Lunar and Cislunar Space. Observations and experiments in the interplanetary medium from lunar and cislunar space might be desirable for both exploration of the particular content of this region, and investigation of the elongated tail of the magnetosphere and of the solar wind on the side of the earth away from the sun.

Continuous observations of the zodiacal light from the Moon would not be disturbed by terrestrial scattering and airglow. These should give evidence whether or not there are really long time variations in the intensity of the zodiacal light. Further, such observations would also permit continuous monitoring of the zodiacal light between the corona and 30° elongation and beyond 90° elongation. If the "Gegenschein" is a phenomenon caused by an enhanced backward scattering intensity in the scattering diagram of interplanetary dust (SIEDENTOPF 1955, WHIPPLE 1955), this could also be confirmed by observations far from the Earth (SIEDENTOPF).

Another possibility is the continuous measurement of the plasma and magnetic fields in the backward region of the magnetosphere and the interplanetary medium. Especially interesting seems to be the possibility of carrying out ion cloud releases by space probes in this region. Observation of these clouds should be advantageous from the lunar surface, since the observation site need not necessarily be in a twilight region as in experiments carried out by our Institute up to now. Furthermore, any small motion of the cloud due to magnetic forces and solar wind can be better observed from a lunar station which is relatively close to the point of release.

Exploration du milieu interplanétaire par des expériences dans l'espace lunaire et cislunaire. Les observations et les expériences portant sur le milieu interplanétaire, faite à partir de l'espace cislunaire ou lunaire sont souhaitables à la fois pour l'exploration du contenu particulier de cette région et pour l'étude de la queue de la magnetosphère et du vent solaire du coté de la terre opposé au soleil.

Les observations continues de la lumière zodiacale à partir de la lune ne seraient pas perturbées par la diffusion terrestre et la lueur atmosphérique. Elles apporteraient la preuve qu'il existe ou non réellement des variations lentes de l'intensité de la lumière zodiacale. De telles observations permettraient de plus un contrôle continu de la lumière zodiacale entre la couronne et 30° d'élongation et au delà de 90° délongation. Si le "Gegenschein" est un phénomène produit par une augmentation de l'intensité de la rétrodiffusion dans le diagramme de diffusion de la poussière interplanétaire (SIEDENTOPF 1955, WHIPPLE 1955) cette interprétation serait confirmée par des observations loin de la terre (SIEDENTOPF).

---

[1] Max-Planck-Institut für Physik und Astrophysik, Institut für extraterrestrische Physik, Garching bei München, German Federal Republic.

Une autre possibilité offerte par le laboratoire est la mesure continue du champ de plasma et du champ magnétique dans la région arrière de la magnétosphère et dans le milieu interplanétaire. La possibilité de libérer des nuages d'ions à l'aide de sondes spatiales dans cette région semble particulièrement intéressante. L'observation de ces nuages à partir de la surface lunaire serait avantageuse puisque le site d'observation ne devrait pas nécessairement être dans une région de crépuscule comme pour les expériences effectuées jusqu'à présent dans nos observatoires. De plus, tout mouvement, même faible, du nuage dû aux forces magnétiques ou au vent solaire serait mieux observé d'une station lunaire, relativement plus proche du point de formation.

Исследования межпланетной среды с помощью экспериментов в лунном и прилунном пространстве. Наблюдения и эксперименты, связанные с изучением межпланетной среды, проводимые в лунном и прилунном пространстве, могут оказаться желательными как для исследования содержания частиц в этом районе, так и для изучения отдаленных областей магнитного поля и солнечных ветров на скрытой от Солнца стороне Земли.

Непрерывным наблюдениям зодиакального света с Луны не будет препятствовать наблюдаемое с Земли рассеяние и свечение ночного неба. Это позволит установить, существуют ли действительно долгосрочные колебания интенсивности зодиакального света. Кроме того, такие наблюдения позволят непрерывно следить за зодиакальным светом в районе между короной и 30° небесной долготы и далее в районе за 90° небесной долготы. Если противосияние вызывается повышением интенсивности обратного рассеяния в диаграмме рассеяния межпланетной пыли (Зидентопф, 1955 г.; Виппл, 1955 г.), то это тоже может быть подтверждено наблюдениями, выполненными на большом удалении от Земли (Зидентопф).

Открывается также возможность непрерывных измерений плазменного и магнитного полей в удаленных районах магнитной сферы и в межпланетном пространстве. Особенно интересными кажутся эксперименты с созданием в этих районах ионных облаков с помощью космических ракет. Наблюдение этих облаков будет удобно выполнять с поверхности Луны, поскольку в этом случае место наблюдений не обязательно должно находиться в сумеречном районе, как это имеет место при проведении экспериментов нашим институтом до сих пор. Более того, любые небольшие движения облака, вызываемые магнитными силами и солнечным ветром, лучше могут быть обнаружены при наблюдении с лунной станции, находящейся относительно ближе к месту выпуска облака, чем земная станция.

1. Introduction

The interplanetary medium has been studied during recent years, both theoretically [1—4] and by the use of space vehicles. There are several reasons for this interest and for the attention given to problems related to interplanetary space. First of all—with the exception of our Earth itself and its atmosphere—the interplanetary medium is the nearest "cosmical object" which can now be investigated by means of space vehicles. Furthermore, it causes or influences markedly a large number of effects observed in the Earth's closest environment such as aurora, magnetic storms and the influx of cosmic rays coming from outer space. The investigation of magnetic fields and interplanetary plasma clouds contributes to understanding of solar activity and its correlation with terrestrial phenomena. Studies of the composition and spatial distribution of interplanetary dust particles gives us more insight into the motion of meteoritic material, its interaction with light and gravitational fields, its supply by comets and planetoids and possibly even into the problem of the origin of our planetary system [5].

Unfortunately, observations of the interplanetary medium from the surface of the Earth are either rather complicated or impossible. Charged particles of low energy are deflected by the magnetic field of the earth. The major part of the solar wind, which is identical with the continuously flowing gas component of the interplanetary medium, cannot penetrate the boundary of the magnetosphere, and is therefore excluded from direct measurement by equipment on the surface of the Earth. Only during the last few years has it been studied directly by space missions. Its velocity was found to be between 300 and 900 km/sec, and, on the average, under undisturbed conditions, 350 km/sec. Its average density is between 1 and 20 particles/cm³ [6—9].

The magnetic field in interplanetary space has recently been measured too, particularly by NESS et al. [10—12]. Its average magnitude was found to be close to 5 gamma. The duration of these measurements, however, was restricted to the time interval when the satellite was in the region of its orbit outside the magnetosphere (e.g. IMP I) or when a space probe had passed the magnetosphere and was still close enough to transmit data by telemetry (e.g. MARINER II).

Information on interplanetary dust has come from observations of meteors, analysis of accretion of cosmic spherules, and from photometry of interplanetary light scattering (F-corona, Zodiacal light, gegenschein). Since BEHR and SIEDENTOPF [13] have shown how observations of the surface brightness I (ε) and the polarization p (ε) of the Zodiacal light can be used to derive models of the spatial distribution of interplanetary dust particles and electrons, many observers have tried to obtain I (ε) and p (ε) as a function of the elongation ε [14—17]. The elongation is the angle between the observer's line of sight and the direction to the sun as measured in the ecliptic plane, close to which most of the measurements were performed. Unfortunately, even the more recent measurements obtained from stations situated on high mountains (e.g. BLACKWELL and INGHAM [18], ELSÄSSER [19], WEINBERG [20] show considerable differences in the resulting functions I (ε) and p (ε). In spite of the laborious care taken in data reduction, these deviations arise from the difficulty of separating the true Zodiacal light from the surface brightness of the sky, caused by star background and especially by terrestrial atmospheric scattering and airglow, which varies with time and direction in an uncontrollable way.

For all these reasons the lunar surface and cislunar space will offer considerable advantages for observing continuously the interplanetary medium, since the Moon is not surrounded by an atmosphere, and since its magnetic field is very small, if it exists at all. In the following a short survey will be given about possible experiments and observations concerning the gas and dust component of the interplanetary medium from lunar and cislunar space.

2. Problems of Interest

a) The Magnetic Field of the Moon

For observations of the solar wind from lunar space the existence or absence of an intrinsic lunar magnetic field is an essential question. The problem has been discussed recently in connection with the magnetohydrodynamic wake of the Moon by N. F. NESS [21], who has detected his phenomenon by magnetic field measurements on the IMP I satellite. The measurements carried out with the Soviet space probe Lunik 2 [22] show that the lunar field must be smaller than 30 gamma at a distance of 55 km from the surface of the moon (30 gamma was the sensitivity of the magnetometer). But as VESTINE [23] and NEUGEBAUER [24] pointed out, the solar wind is sufficiently strong to confine the magnetic field of the moon to a region very close to its surface. In Fig. 1 (NESS, 1965) the radius R_C of the lunar magnetic cavity boundary at the subsolar stagnation point (in units of the lunar radius R_M) is given as a function of the solar wind flux and the equatorial field strength (B_0) of an assumed lunar dipole.

But even if the moon has no inherent magnetic field, the solar wind would not hit the surface of the moon, as Gold [25] has pointed out, due to the interaction of the magnetized solar plasma with the lunar body. The interplanetary magnetic field will be trapped by the surface layers of the moon due to the finite electrical conductivity of the moon. The field strength, which will be built up on the front surface will equal the stagnation field strength as above. For a normal solar wind (e.g. 4 particles/cm³, 300 km/sec) the stagnation field is 30 gamma. The enhanced

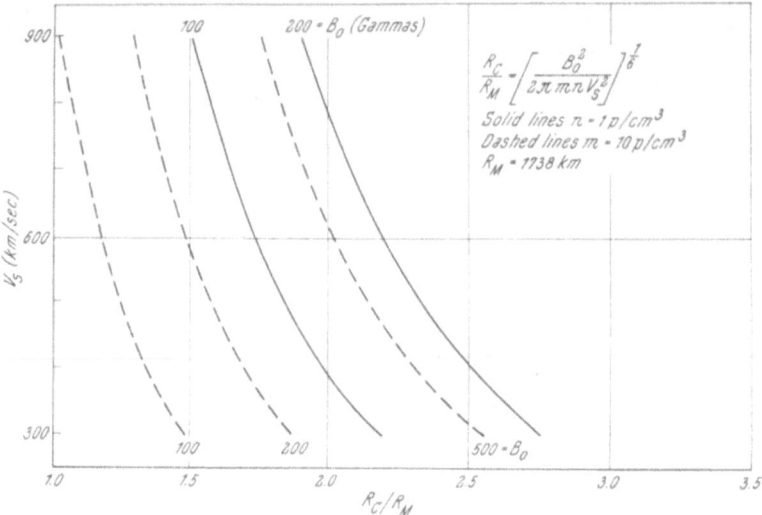

Fig. 1. Lunar magnetic cavity radius of curvature (R_C) in lunar radii (R_M) as a function of the solar wind flux and equatorial field strength (B_0) of assumed lunar dipole (Ness 1965)

field will be quite close to the front surface and the thickness of the magnetosphere of the moon facing the sun will be, according to Gold, only a few kilometers. It would be valuable to confirm or to improve these conclusions by measurements near and above the surface of the moon.

b) Existence of a Collisionless Shock Wave

Even outside the thin magnetic skin or outside of the magnetic cavity one might not find the free-streaming solar wind since a collisionless magnetohydro-dynamic shock wave might enclose the lunar magnetosphere, as in the case of the Earth's magnetosphere. The shock front will exist only if its thickness is much less than the Moon's radius. The thickness will be of the order of the radius of gyration of the solar wind particles. In the relatively weak interplanetary magnetic field this radius will be for a solar wind proton comparable with the lunar radius: for a 500 eV proton (300 km/sec) in a magnetic field of 5 gamma we arrive at a radius of gyration of 646 km. For this reason the development of such a collision-less shock wave depends very critically on the state of the interplanetary medium and on the state of the lunar magnetic field. Therefore, it would be very inter-esting to carry out observations in the region where one might expect such a shock wave and we may hope to observe its development in time, if the shock wave does not always exist. The stand-off distance R_S of the shock wave might be about 500 km at the stagnation point, if the figure $R_S/R_C = 1.30$ is the same as in the case of the earth.

c) Solar Wind Measurements

From the arguments referred to above it is evident that one would not observe the solar wind directly by instruments located on the surface of the Moon. But a satellite in an orbit close to the lunar surface might carry out valuable measurements concerning the solar wind and its interaction with a planetary object. A further object of interest for exploration by a lunar satellite could be the tail of the Earth's magnetosphere. Its radius has been found to be about 20 Re (Earth's radii) in a distance of 30 Re by the IMP I measurements (see NESS [26]). If the radius would be in lunar distance of the order of 20 Re, the moon would be crossing the tail (see Fig. 2) for about 3 days per month.

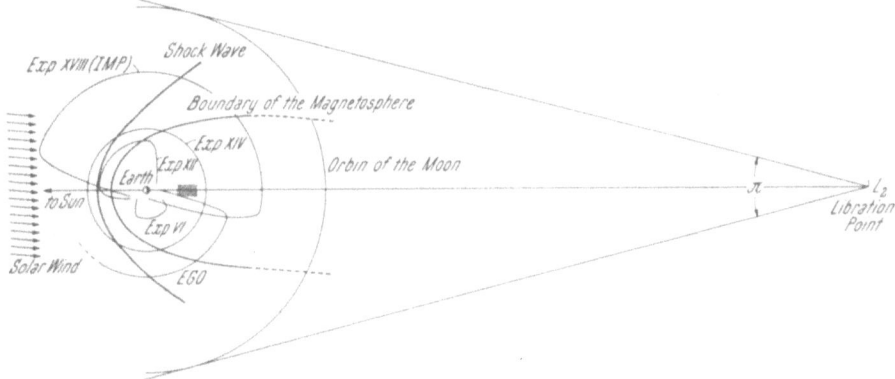

Fig. 2. Schematic diagram of the orbit of the Moon around the Earth, the boundary of the magnetospheric regions observed from several satellites and hypothetical locations of the particles producing gegenschein

Due to the inclination of the lunar orbit the crossings would take place roughly between 5.4 Re above and 5.4 Re below the ecliptic plane, and complete a rather instructive set of scannings across this region within the course of one year. On the other hand, the Moon will be outside the magnetospheric tail for about 26 days per month and would allow continuous monitoring of the undisturbed solar wind. Additional information may be obtained if the orbit of the lunar satellite crosses the region where the solar wind is shielded by the Moon itself. This would give repeatedly the opportunity of studying the interaction between the solar wind and an object having a very small magnetic field or no field at all, like the comets [27, 59, 60].

Another opportunity of studying the behaviour of the solar wind is the possibility of carrying out ion-cloud experiments in lunar and cislunar space. It has been proposed [28—30] that these experiments be carried out in interplanetary space and be observed from the Earth. The most suitable elements for such experiments are the heavier alkaline earth metals and some of the rare earths. At present we are planning to use Barium, since both the neutral atom and its ion have strong resonance lines in the visible spectral region, and the probability of photo-ionization by sunlight seems to be quite high. Several experiments in the upper atmosphere have been carried out [31] so far to investigate the feasibility of this method for studying the state of the interplanetary medium and learning more about the interaction between the solar wind and cometary tails. In these experiments ion-clouds have been observed which were elongated in the direction of the field lines of the Earth's magnetic field in contrast to the spherical neutral clouds.

If the ion cloud technique proves to be capable of producing measurable interactions with the interplanetary plasma in our next experiments, investigations of the advantages of a lunar observing site should be kept in mind. The ion-clouds could be released in interplanetary space much closer to the surface of the Moon as compared to the surface of the Earth. Therefore any spatial shift of the cloud due to the acceleration by the solar wind would be much easier to measure from the Moon than from the surface of the Earth. Furthermore, experiments may be performed even on the illuminated side of the Moon and in less restricted spectral regions, because of the absence of a lunar atmosphere.

d) The Gegenschein

An interesting question which may be solved by observations from the Moon or from lunar space is the origin of the gegenschein. This enhanced and possibly varying [32] brightness [33] near the antisolar point of the sky has been understood as an effect caused by the light scattering function of dielectric dust particles [34—37]; as a dust accretion near the libration point L_2 of the Sun-Earth system (GYLDEN, MOULTON [38]); or as a comet-like tail of the Earth composed of dust, gas or plasma [39—41]. Since none of these theories alone offers a simple and completely satisfying interpretation of the gegenschein (see [37, 42]), it is highly desirable to remove these uncertainties by observations from a highly excentric orbit (SIEDENTOPF ESRO proposal S 19) or from lunar space. Fig. 2 shows schematically the orbit of the Moon around the Earth as projected onto the ecliptic plane. To give an impression of scale and orientation, the boundary of the magnetosphere, the direction to the Sun, and the region observed by means of several artificial satellites are also shown in the Figure. The shaded area designates the region where the particles which cause the gegenschein should be located, if the diurnial parallax of the gegenschein [43] is taken as a reality. L_2 is the libration point where the particle cloud, assumed from the theory of GYLDEN and MOULTON, should be found — although it is not very likely that such a cloud, having 100 times the particle density of the interplanetary medium, exists at this point [44, 45].

As can be seen from Fig. 2, even a particle accretion near the comparatively distant point L_2 would show a parallax of about 30° when observed from opposite regions of the Moon's orbit. Any cloud of sufficient brightness inside the orbit or a tail crossing the lunar orbit could not escape identification by its parallax. If, however, the gegenschein is not based on a localized dust cloud or tail, but on a general effect of enhanced backward scattering of the interplanetary dust, there should be no changes in brightness and direction of the gegenschein when the phenomenon is observed from different parts of the lunar orbit.

e) The Zodiacal Light

How much the lack of reliability of data gained by Earth-based observations of the zodiacal light may affect an interpretation of the interplanetary medium can be seen from Fig. 3. It shows the curves $I(\varepsilon)$ and $p(\varepsilon)$ resulting from a theoretical model of interplanetary matter assuming spherical particles of both dielectric and absorbing material in size between $3 \cdot 10^{-5}$ and $3 \cdot 10^{-3}$ cm, and having a differential distribution function of the particle radii (r) proportional to $r^{-2.5}$, which is compatible with the distribution functions derived by VAN DE HULST [46] and ELSÄSSER [47] from the F-corona. Comparison between the theoretical curve (B) and the measurements of WEINBERG (A) shows, that such a model might approximate the observations exclusively by particles of the composition as outlined above. Contrary to this result it seems not to be possible to approximate BLACKWELL's (●) and ELSÄSSER's (○) data by a model of spherical particles having a similar size distribution. Although such dielectric and absorbing particles are able to cause high polarization (25% or 40% respectively) it is necessary at least within the restriction of a spherical particle model to assume

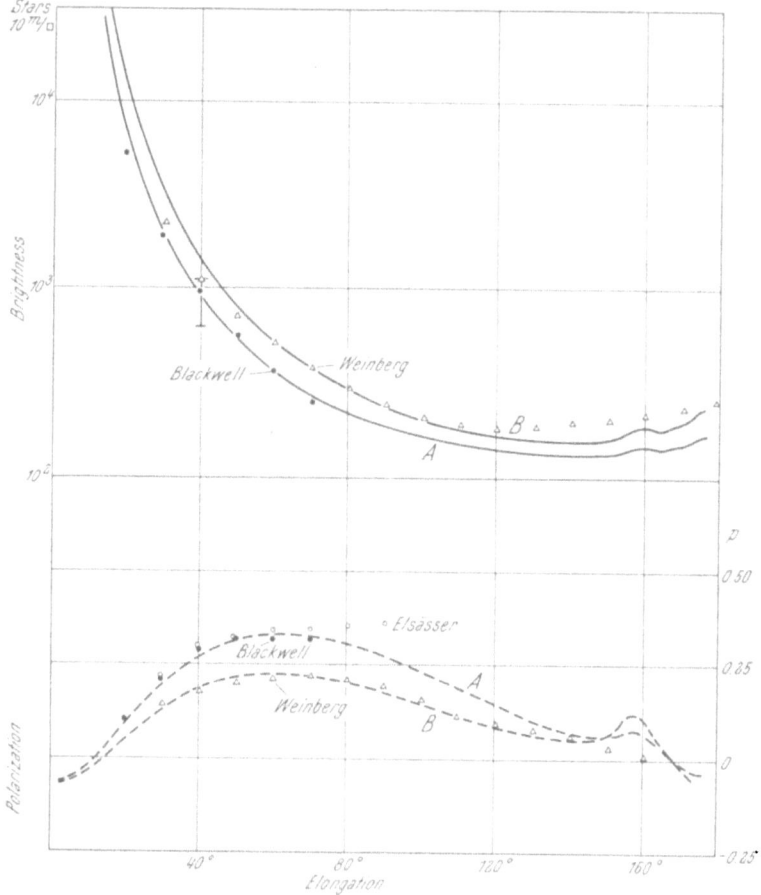

Fig. 3. Theoretical models approximating the Zodiacal light observations of BLACKWELL, ELSÄSSER and WEINBERG. A: Model (B 1) of GIESE and SIEDENTOPF (Z. Astrophysik 54, 213, 1962); B: Same dust composition as Model (B 1), $2.7 \cdot 10^{-13}$ particles/cm³ near Earth's orbit, no electrons

an additional component either of electrons (THOMPSON scattering, about 300—500 cm⁻³ near the earth's orbit [48]) or of very small particles (RAYLEIGH scattering [49]) in order to reproduce the shape of $p(\varepsilon)$ with the observed polarization rising to about 35% in the region above 60° of elongation. To resolve such ambiguities more reliable, extraterrestrial measurements are needed in this range of elongation. In addition, photometry of the zodiacal light from space vehicles or from a lunar observatory would permit observation of the regions of elongation $\varepsilon < 20°$, between $90° < \varepsilon < 180°$, and far from the ecliptic plane. In these regions observations are very seriously disturbed by the atmosphere, and, therefore, only a few sets of data are available up to now (e.g. [20, 50—52]). Observations at small elongations will give more evidence about the true size distribution of interplanetary dust particles, since forward scattering is highly dependent on the size of the diffracting bodies. On the other hand, observations in the region of 120° may give some insight into the abundance of dielectric and absorbing material [53]. Photometry far from the ecliptic plane is valuable for deriving a three dimensional representation of dust distribution in our solar system.

While the measurements mentioned above can be made from appropriate regular earth satellites, monitoring of possible fluctuations of the zodiacal light requires a permanent, extraterrestrial station which could well be located on the Moon. Because of fluctuations of terrestrial airglow, it is not possible to decide beyond any doubt if there remain real fluctuations of the zodiacal light after reduction of the photometric data (see [20, 54]). However, BLACKWELL and INGHAM seem to have found an enhancement of brightness after a solar flare [54]), and even a relation between the zodiacal brightness and the lunar phase has been reported in recent literature [55]. As a very tentative interpretation of their observation (at $\lambda = 6{,}200$ Å and $4{,}500$ Å) after a solar flare, BLACKWELL and INGHAM consider an ejected cloud of solar plasma having constant particle density (300 cm^{-3}), and contributing, by THOMPSON scattering, to the brightness observed at $\varepsilon = 30°$ along a path length of 0.5 astronomical units [54].

Similar figures necessary to cause a measurable increase in brightness and polarization of zodiacal light can be found from our earlier computation using a number density of electrons proportional to $\varrho^{-\beta}$ (with ϱ the distance from the Sun, in astronomical units [48]). For an additional electron component of 1,000 or 100 cm^{-3} near the Earth's orbit, using the exponent $\beta = 1.5$, we obtain an increase of zodiacal light from 2,200 (WEINBERG) to 3,100 or from 2,200 to 2,290 stars of 10th magnitude per square degree. At the same time the polarization would rise from 14.8 to 27.2% or from 14.8 to 16.5%, respectively. These figures show that plasma streams of similar density ranging over path lengths of the order of 1 astronomical unit might be observable by extraterrestrial photometry.

Unfortunately, there is not much hope of observing scattered light from plasma clouds of small size (10^4—10^5 km) and moderate particle density (some 10 particles/cm^3). Nevertheless, monitoring the zodiacal light will be of great importance, especially since the mechanism producing the fluctuations, which were reported up to now, is still open for discussion.

f) Recovery of Interplanetary Particles

In addition to optical measurements, direct capture of interplanetary dust particles in the lunar and cislunar space should be of great value. According to WHIPPLE [56], the Earth is surrounded by a dust cloud which decreases proportional to a $R^{-1.4}$ law, that is, from 10^4 times the interplanetary particle density (near the earth) to the density derived from the zodiacal light (10^5 km from the Earth), if R is the distance from the Earth's surface. Therefore, the samples of extraterrestrial dust obtained by SOBERMANN [57] might not represent in size and shape exactly the properties of interplanetary particles. Especially, there remains the questions: are the submicron size particles, which would cause RAYLEIGH scattering, present in the more distant regions of interplanetary space or are they generated near the Earth by crushing of larger, fluffy particles, which were also found by SOBERMANN's sounding rocket experiments. In the same way, it is not clear if the light scattering function that was adopted by FESENKOV [58] from scattering measurements in the higher atmosphere is representative for interplanetary particles in order to derive the interplanetary dust distribution.

Samples of particles from the lunar and cislunar space which may be analyzed by a lunar laboratory may answer the questions: are there really differences; does the supply of the terrestrial dust cloud come from lunar dust (see [45]); and are optical models favouring submicron particles (see [49], [20], [53]) realistic enough to represent interplanetary dust scattering, and to draw from this kind of optical models significant conclusions concerning the distribution of interplanetary dust grains in more distant regions of the solar system.

3. Conclusion

Measurements near the Moon are necessary for the exploration of the interaction between the Moon and the solar wind. Furthermore, experiments and observations in lunar and cislunar space are useful for investigating the Earth's magnetic tail, the nature of the gegenschein and of interplanetary dust. Although we feel that a Moon-based station is not absolutely necessary for these investigations, a lunar international laboratory would at least be very useful as a platform, especially for long time observations, such as needed for monitoring fluctuations of the zodiacal light.

References

1. E. N. PARKER, Planet. Spac. Sci. 9, 461 (1962).
2. R. LÜST, Spac. Sci. Rev. 1, 522 (1962).
3. J. H. PIDDINGTON, Planet. Spac. Sci. 6, 305 (1962).
4. J. H. PIDDINGTON, Planet. Spac. Sci. 13, 363 (1965).
5. F. L. WHIPPLE, Publ. Astronom. Soc. Pacific 70, 485 (1958).
6. A. BONETTI, H. S. BRIDGE, A. S. LAZARUS, B. ROSSI and F. SCHERB, J. Geophys. Res. 68, 4017 (1963).
7. C. SNYDER and M. NEUGEBAUER, Spac. Res. 4, 89 (1964).
8. H. BRIDGE, A. LAZARUS, E. LYON and L. JACOBSON, Spac. Res. 5, 969 (1965).
9. K. I. GRINGAUZ, Spac. Res. 2, 539 (1961).
10. N. F. NESS, G. S. SCEARCE and J. B. SEEK, J. Geophys. Res. 69, 3531 (1964).
11. C. P. SONETT, E. J. SMITH and A. R. SIMS, Spac. Res. 1, 921 (1960).
12. P. J. COLEMAN, L. DAVIS and E. J. SMITH, JPL Symposium Solar Wind, Pasadena, Calif. (1964).
13. A. BEHR and H. SIEDENTOPF, Z. Astrophys. 32, 19 (1953).
14. D. BARBIER, Mem. Soc. Roy. Sci. Liège 15 (1955).
15. N. B. DIVARI and A. S. ASAAD, Soviet Astronomy A. J. 3, 832 (1960).
16. A. W. PETERSON, Astronom. J. 133, 668 (1961).
17. R. ROBEY, Ann. Geophys. 18, 341 (1962).
18. D. E. BLACKWELL and M. F. INGHAM, Monthly Notices Roy. Astronom. Soc. 122, 113 (1961).
19. H. ELSÄSSER, Die Sterne 9—10, 166 (1958).
20. J. L. WEINBERG, Ph. D. Thesis University of Colorado (1963).
21. N. F. NESS, J. Geophys. Res. 70, 517 (1965).
22. S. SH. DOLGINOV, E. G. EROSHENKO, L. N. ZHUGOV, N. V. PUSHKOV and L. O. TYURMINA, Artificial Earth Satellites 5, 490 (1961). (Translation from Russian.)
23. E. H. VESTINE, Rand Rep. R. M. — 1933 (1959).
24. M. NEUGEBAUER, Phys. Rev. Letters 4 (1), 6 (1960).
25. T. GOLD, Cornell Univ. Rept. TSR-177 (1964).
26. N. F. NESS, J. Geophys. Res. 70, 2989 (1965).
27. L. BIERMANN, Z. Astrophys. 29, 274 (1951).
28. L. BIERMANN, R. LÜST, RH. LÜST and H. H. SCHMIDT, Z. Astrophys. 53, 226 (1961).
29. E. R. HARRISON, Geophys. J. 6, 462 (1962).

30. E. W. HONES, Private Communication.
31. H. FÖPPL, G. HAERENDEL, J. LOIDL, R. LÜST, F. MELZNER, B. MEYER, H. NEUSS and E. RIEGER, Planet. Spac. Sci. 13, 25 (1965).
32. L. M. GINDILIS, Soviet Astronomy A. J. 6, 67 (1962).
33. H. ELSÄSSER and H. SIEDENTOPF, Z. Astrophys. 43, 132 (1957).
34. H. SIEDENTOPF, Z. Astrophys. 36, 240 (1955).
35. H. WALTER, Z. Astrophys. 46, 9 (1958).
36. F. L. WHIPPLE, Trans. I.A.U. 9, 321 (1955).
37. L. M. GINDILIS, Soviet Astronomy A. J. 6, 540 (1963).
38. MOULTON, Celest. Mechanics.
39. V. G. FESENKOV, Astronom. Zhur. 27, 89 (1950).
40. V. G. FESENKOV, Soviet Astronomy A. J. 8, 803 (1965).
41. J. H. PIDDINGTON, J. Geophys. Res. 65, 93 (1960).
42. M. F. INGHAM, Spac. Sci. Rev. 1, 576 (1962/1963).
43. D. A. ROZHLOVSKII, Astronom. Zh. 27, 34 (1950).
44. U. HAUG, Diplomarbeit, Tübingen (1964).
45. G. COLOMBO, D. LAUTMANN and I. SHAPIRO, Proc. Symposium on Micrometeorites and Cosmic Dust at the Max-Planck-Institut f. Kernphysik, Heidelberg (1965).
46. H. C. VAN DE HULST, Astrophys. J. 105, 471 (1947).
47. H. ELSÄSSER, Z. Astrophys. 37, 114 (1955).
48. R. H. GIESE and H. SIEDENTOPF, Z. Astrophys. 54, 200 (1962).
49. R. H. GIESE, Spac. Sci. Rev. 1, 589 (1962/1963).
50. D. E. BLACKWELL, Monthly Notices Roy. Astronom. Soc. 115, 629 (1955).
51. D. W. BEGGS, D. E. BLACKWELL, D. W. DEWHIRST and R. D. WOLSTENCROFT, Monthly Notices Roy. Astronom. Soc. 128, 419 (1964).
52. L. L. SMITH, F. E. ROACH and R. W. OWEN, Planet. Spac. Sci. 13, 207 (1965).
53. S. J. LITTLE, B. J. O'MARA and L. H. ALLER, Astronom. J. 70, 346 (1965).
54. D. E. BLACKWELL and M. F. INGHAM, Monthly Notices Roy. Astronom. Soc. 122, 143 (1961).
55. N. B. DIRAVI, Soviet Astronom. A. J. 7, 547 (1964).
56. F. L. WHIPPLE, American Rocket Society, Space Flight Report to the Nation, New York 1961; Nature 189, 127 (1961).
57. R. K. SOBERMANN (ed.), GRD Research Notes No. 71, Bedford, Mass. (1961).
58. V. G. FESENKOV, Soviet Astronomy A. J. 2, 475 (1958).
59. L. S. MAROCHNIK, Geomagnetism and Aeronomy 3, 608 (1963).
60. H. ALFVÉN, Tellus 9, 92 (1957).

Observations of Star Fields Made Beyond the Atmosphere

By

N. A. Dimov[1] and A. B. Severny[2]

(With 3 Figures)

Abstract — Résumé — Резюме

Observations of Star Fields Made Beyond the Atmosphere. It is important in astrophysics to measure the brightness of comparatively large star fields, taking measurements simultaneously in the ultraviolet and the visible part of the spectrum beyond the Earth's atmosphere.

From measurements of the sky background it is also possible to draw conclusions about the brightness of the interplanetary dust, and the possibility of discovering stars and nebulae too faint to be observed from the Earth.

Such measurements can be made by means of a wide-angled photoelectric photometer, with a field of vision of approximately 0.1 steradian.

We made the first experiment by installing, on the artificial satellite "Kosmos-51", a photometer with an angle of vision of $\sim 20°$ measuring the brightness of the sky in three parts of the spectrum — 3700—5500 Å, 2200—3000 Å and 1260—1300 Å. Measurements of this last named part of the spectrum were recorded during the first few revolutions only, measurements of the first two parts for the duration of a month.

The following provisional deductions can be made:

Firstly, the smallest value for the brightness of the sky, according to measurements taken in the visible part of the spectrum, is approximately 30% of the brightness observed from the Earth — in other words, the value is rather higher than was expected. It should be noted in this connexion that for measurements during flight, calibrations were made fairly frequently from a radioactive luminophore.

Secondly, there are substantial fluctuations of the ultraviolet color-index in some parts of the sky.

Observation des champs stellaires hors de l'atmosphère. La possibilité de faire, hors de l'atmosphère terrestre, et simultanément dans le visible et l'ultraviolet, la mesure de la luminance de champs stellaires relativement grands présente de l'intérêt pour l'astrophysique.

Les mesures sur le fond du ciel permettent de tirer des conclusions sur la luminance des poussières interplanétaires et sur les possibilités de découvrir des étoiles et des nébuleuses plus faibles que celles observables sur terre.

La réalisation des ces mesures est possible utilisant un photomètre photoélectrique de très grand angle, dont le champ d'ouverture est de l'ordre de 0,1 stéradian.

Comme première expérience, nous avons installé sur le satellite artificiel de la terre Kosmos 51 un photomètre d'angle d'ouverture "environ 20°" mesurant la luminance du ciel dans trois domaines du spectre — 3700—5500 Å, 2200—3000 Å et

---

[1] Crimean Observatory, Nauchny, Crimea, U.S.S.R.
[2] Director, Crimean Observatory, Nauchny, Crimea, U.S.S.R.

1260—1300 Å. Le dernier domaine du spectre a été enregistré pendant les premières révolutions, les autres pendant un mois.

Les résultats obtenus permettent de donner les conclusions suivantes:

Permièrement, la luminance minimale du ciel, d'après les mesures dans la partie visible du spectre, représente environ 30% de la luminance qu'on obtient de la terre, c'est-à-dire qu'elle excède un peu celle qu'on peut attendre. On peut mentionner ici que les mesures en cours de vol s'accompagnèrent de réglages assez fréquents à l'aide d'un photophore radioactif.

Deuxièmement, on observe des fluctuations importantes de l'indice de couleur ultraviolet dans certaines régions du ciel.

Внеатмосферные наблюдения звездных полей. Для астрофизики представляет интерес измерение светимости сравнительно больших звездных полей, проводимое одновременно в ультрафиолетовой и видимой части спектра вне атмосферы Земли.

Измерение фона неба позволяет также сделать заключение о светимости межпланетной пылевой составляющей и о возможности обнаружения звезд и туманностей более слабых чем наблюдаемые с Земли.

Осуществление таких измерений возможно при использовании даже весьма широко-угольного фотоэлектрического фотометра с полем зрения порядка 0,1 стерадиана.

В качестве первого опыта нами был установлен на искусственном спутнике Земли „Космос-51" фотометр с углом зрения ~20°, измеряющий светимость неба в трех областях спектра 3.700—5.500 A, 2.200—3.000 A и 1.260—1.300 A. Последняя область спектра регистрировалась в течение нескольких первых витков, а остальные в течение месяца.

Полученные данные позволяют сделать предварительно следующие заключения: во-первых, наименьшая светимость неба по измерениям в видимой части спектра составляет около 30 проц. светимости, наблюдаемой с Земли, т.е. несколько превышает ожидаемую. Отметим здесь, что измерения в полете проводились с достаточно частой калибровкой от радиоактивного люминофора, который в свою очередь был откалиброван по свечению ночного неба с Земли.

Во-вторых, наблюдаются значительные флуктуации ультрафиолетового колор-индекса в некоторых областях неба.

1. The usual brightness measurement of the night sky as taken from the Earth includes a space component and an Earth component. The space component is due to stars and to interplanetary and galactic dust. The Earth component is produced by emission of the upper layers of our atmosphere (consisting of single emission lines and bands of oxygen, sodium, nitrogen, hydroxyl and of continuous emission), and by scattering of radiation of all the above mentioned sources. Only the stellar component can be evaluated reliably (by counting the stars of different magnitudes), the emission in single lines and bands can also be estimated with a certain degree of accuracy, but the other components, and in particular those belonging to interstellar dust and gas, cannot be evaluated directly.

2. The measurements of night sky brightness made outside the Earth's atmosphere give us the possibility to exclude completely the influence of the radiations of the Earth's atmosphere. Further, they allow us to determine the dust and gas components directly from the known numbers of stars of different magnitudes. At the same time we get an idea about the limiting magnitudes or on the penetrating powers of telescopes operating on the Moon or in space. Making these measurements one should use several spectral regions, if one wishes to obtain information on the "color" of the radiation observed or about the possibilities of making observations in the ultraviolet region.

3. The measuring apparatus used was a three-channel photoelectric photometer, AF-3. Measurements in the visible part of the spectrum were made by

means of a FEU-15 type bismuth-cesium photomultiplier; those in the ultraviolet part of the spectrum, by means of a FEU-57 photomultiplier with maximum sensitivity in the region of 2800 Å, and a photon counter sensitive in the region of 1225—1340 Å. Unfortunately, the photon counter operated only during the first revolutions, and the information obtained was insufficient for processing. Fig. 1 gives a block diagram of the photometer. In addition to the rotating modu-

lator M which, in order to reduce the influence of the dark current, modulates the light falling on the photomultipliers with a frequency of 120 c.p.s., there is also a programme disk P, which places various diaphragms in front of the cathodes of the photomultipliers, thus reducing their working surfaces by a known proportion. This enables one to widen the range of intensities measured. Once every 10 seconds the disk P sets a radioactive light standard before the cathode of the FEU-15 for about 2 seconds, in order to control the instrument's sensitivity in the visible region of the

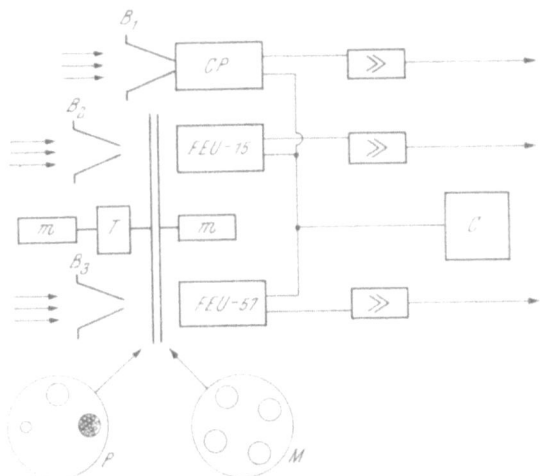

Fig. 1. Block diagram of the photometer AF-3

spectrum. The diaphragms $B_1 — B_3$ limit the field measured to 18° for all receptors. Special care has been devoted to the design and position of these diaphragms in order to reduce the influence of scattered light (with an angle of incidence of more than

70°, the amount of scattered light is less than 10^{-5}). Photomultipliers and a. c. amplifiers are fed from converter C. Fig. 2 shows a general view of the AF-3, while Table 1 contains data relating to the sensitivity of the photomultipliers, and to the flux measured from the radioactive standard.

4. The photometer AF-3 was installed on the artificial satellite "Kosmos 51", launched

Fig. 2. General view of the AF-3 photoelectric photometer

on 10 December, 1964. The bulk of data was obtained during December 1964. As "Kosmos 51" was not stabilized, the identification of the fluxes measured and of the areas in the sky is a difficult problem. This work is now being done. Preliminary reductions of telemetric records have been carried through for the cases when the photometer was "looking" at the sky while Sun as well as Moon were absent.

Table I

FEU-15	3200—6800 Å	10^{-12}—$5 \cdot 10^{-11}$watt
FEU-57	2200—3400 Å	7.10^{-12}— 10^{-10}watt
Standard	3800—6000 Å	10^{-11}watt

In Fig. 3 the top curve represents a record in the visible part of the spectrum expressed in units of the flux from the radioactive standard, the corresponding record in the ultra-violet (2800 Å) is represented by the bottom curve. The middle curve shows the ratio between the records in the visible and the ultra-violet region. The time of recording is marked on the abscissa, and the recorded flux, in watts, on the ordinate. The examination of many similar records showed that the mini-

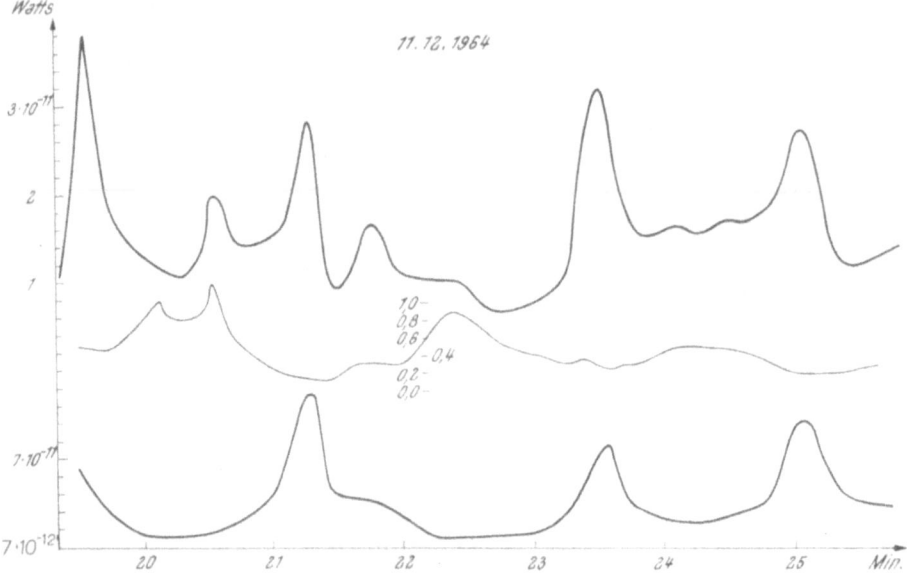

Fig. 3. Top curve — visible part of the spectrum in relation to the light standard; bottom curve — ultraviolet region (2800 Å); middle curve — ratio between these curves

mum night sky brightness as observed outside the Earth's atmosphere is equivalent to 190 stars of magnitude 10 per square degree, i. e., twice the amount expected on the basis of simply counting the stars of different magnitudes in middle galactic latitudes [1].

Remarkable fluctuations of the ratio of the brightness of the night sky in the visible to that in the ultra-violet part of the spectrum have been observed too. These fluctuations exceed by far those expected from the calculations of the abundances of stars of different magnitudes and spectral classes.

The authors are indebted to Mr. V. V. Beniuh who cooperated in constructing and testing the photometer. The assistence of Mr. A. P. Kulchizky who mounted the electronic circuits is highly appreciated.

Reference

1. C. W. Allen, Astrophys. Quantities 125, 1955.

Druck: Globus, Wien XX